MVFOL

# The ESSENTIALS® of

# Algebra &
# Trigonometry I

D0190574

**Staff of Research and Education Association**
**Dr. M. Fogiel, Director**

This book covers the usual course outline of
Algebra and Trigonometry I. For more advanced
topics, see *"THE ESSENTIALS OF ALGEBRA &
TRIGONOMETRY II."*

*Research & Education Association*
61 Ethel Road West
Piscataway, New Jersey 08854

# THE ESSENTIALS®
## OF ALGEBRA & TRIGONOMETRY I

Year 2002 Printing

Printed in the United States of America

Library of Congress Control Number 00-134281

International Standard Book Number 0-87891-569-9

# WHAT "THE ESSENTIALS" WILL DO FOR YOU

This book is a review and study guide. It is comprehensive and it is concise.

It helps in preparing for exams and in doing homework, and remains a handy reference source at all times.

It condenses the vast amount of detail characteristic of the subject matter and summarizes the **essentials** of the field.

It will thus save hours of study and preparation time.

The book provides quick access to the important facts, principles, theorems, concepts, and equations in the field.

Materials needed for exams can be reviewed in summary form – eliminating the need to read and re-read many pages of textbook and class notes. The summaries will even tend to bring detail to mind that had been previously read or noted.

This "ESSENTIALS" book has been carefully prepared by experts in the field and has been carefully reviewed to ensure its accuracy and maximum usefulness.

Dr. Max Fogiel
Program Director

# CONTENTS

# 4 POLYNOMIALS AND RATIONAL EXPRESSIONS

# 5 EQUATIONS

# 6 LINEAR EQUATIONS AND SYSTEMS OF LINEAR EQUATIONS

# 7 INEQUALITIES

# CHAPTER 1

# SETS AND SET OPERATIONS

## 1.1 SETS

A set is defined as a collection of items. Each individual item belonging to a set is called an element or member of that set. Sets are usually represented by capital letters, elements by lowercase letters. If an item k belongs to a set A, we write k ε A ("k is an element of A"). If k is not in A, we write k ∉ A ("k is not an element of A"). The order of the elements in a set does not matter:

$$\{1,2,3\} = \{3,2,1\} = \{1,3,2\}, \text{ etc.}$$

A set can be described in two ways: 1) it can be listed element by element, or 2) a rule characterizing the elements in a set can be formulated. For example, given the set A of the whole numbers starting with 1 and ending with 9, we can describe it either as A = { 1,2,3,4,5,6,7,8,9 } or as { the set of whole numbers greater than 0 and less than 10}. In both methods, the description is enclosed in brackets. A kind of shorthand is often used for the second method of set description; instead of writing out a complete sentence in between the brackets, we write instead

$$A = \{k \,|\, 0 < k < 10, \text{ k a whole number}\}$$

This is read as "the set of all elements k such that k is greater than 0 and less than 10, where k is a whole number."

A set not containing any members is called the empty or null set. It is written either as $\phi$ or $\{\ \}$.

# 1.2 SUBSETS

Given two sets A and B, A is said to be a subset of B if every member of set A is also a member of set B. A is a <u>proper</u> subset of B if B contains at least one element not in A. We write $A \subseteq B$ if A is a subset of B, and A $\subset B$ if A is a proper subset of B.

Two sets are equal if they have exactly the same elements; in addition, if A = B then $A \subseteq B$ and $B \subseteq A$.

e.g.  Let    A = $\{1,2,3,4,5\}$

B = $\{1,2\}$

C = $\{1,4,2,3,5\}$

Then 1) A equals C, and A and C are subsets of each other, but not proper subsets and 2) $B \subseteq A$, $B \subseteq C$, B $\subset A$, $B \subset C$ (B is a subset of both A and C. In particular, B is a proper subset of A and C.)

A universal set U is a set from which other sets draw their members. If A is a subset of U then the complement of A, denoted A' or $A^0$, is the set of all elements in the universal set that are not elements of A.

e.g.    If U = $\{1,2,3,4,5,6,\ldots\}$    and    A = $\{1,2,3\}$, then A' = $\{4,5,6,\ldots\}$.

Figure 1.1 illustrates this concept through the use of a <u>Venn</u> <u>diagram</u>.

Fig. 1.1

2

# 1.3 UNION AND INTERSECTION OF SETS

The union of two sets A and B, denoted A ∪ B, is the set of all elements that are either in A or B or both.

The intersection of two sets A and B, denoted A ∩ B, is the set of all elements that belong to both A and B.

If A = {1,2,3,4,5} and B = {2,3,4,5,6} then A ∪ B = {1,2,3,4,5,6} and A ∩ B = {2,3,4,5}.

If A ∩ B = φ, A and B are disjoint. Fig. 1.2 and 1.3 are Venn diagrams for union and intersection. The shaded areas represent the given operation.

A ∪ B
Fig. 1.2

A ∩ B
Fig. 1.3

# 1.4 LAWS OF SET OPERATIONS

If U is the universal set and A is any subset of U, then the following hold for union, intersection, and complement:

## IDENTITY LAWS

1a. $A \cup \phi = A$       1b. $A \cap \phi = \phi$
2a. $A \cup U = U$       2b. $A \cap U = A$

## IDEMPOTENT LAWS

3a. $A \cup A = A$       3b. $A \cap A = A$

## COMPLEMENT LAWS

4a. $A \cup A' = U$       4b. $A \cap A' = \phi$
5a. $(A')' = A$       5b. $\phi' = U;\ U' = \phi$

## COMMUTATIVE LAWS

6a. $A \cup B = B \cup A$       6b. $A \cap B = B \cap A$

## ASSOCIATIVE LAWS

7a. $(A \cup B) \cup C = A \cup (B \cup C)$
7b. $(A \cap B) \cap C = A \cap (B \cap C)$

## DISTRIBUTIVE LAWS

8a. $A \cup (B \cap C) = (A \cup B) \cap (A \cup C)$
8b. $A \cap (B \cup C) = (A \cap B) \cup (A \cap C)$

## DE MORGAN'S LAWS

9a. $(A \cup B)' = A' \cap B'$       9b. $(A \cap B)' = A' \cup B'$

# CHAPTER 2

# NUMBER SYSTEMS AND FUNDAMENTAL ALGEBRAIC LAWS AND OPERATIONS

## 2.1 NUMBER SYSTEMS

Most of the numbers used in algebra belong to a set called the real numbers, or reals. This set denoted ℝ, can be represented graphically by the real number line.

Given a straight horizontal line extending continuously in both directions, we arbitrarily fix a point and label it with the number 0. In a similar manner, we can label any point on the line with one of the real numbers, depending on its position relative to 0. Numbers to the right of zero are called positive, while those to the left are called negative. Value increases from left to right, so that if a is to the right of b, it is said to be greater than b.

Figure 2.1

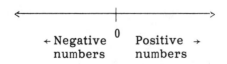

### Integers

If we divide the number line into equal segments called unit lengths, we can then label the boundary points of these segments according to their distance from zero. For example, the point 2 lengths to the left of zero is -2, while the point 3 lengths to the right of zero is +3 (the + sign is usually assumed, so +3 is written as 3). The number line now looks like this:

Figure 2.2

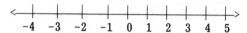

$$-4 \quad -3 \quad -2 \quad -1 \quad 0 \quad 1 \quad 2 \quad 3 \quad 4 \quad 5$$

These boundary points represent the subset of the reals known as the integers, denoted $\mathbb{Z}$. Some subsets of $\mathbb{Z}$ are the natural numbers or positive integers, the set of integers starting with 1 and increasing, $\mathbb{Z}^+ = \mathbb{N} = \{1,2,3,4,\ldots\}$; the whole numbers, the set of integers starting with 0 and increasing, $\mathbb{W} = \{0,1,2,3,\ldots\}$; the negative integers, the set of integers starting with -1 and decreasing: $\mathbb{Z}^- = \{-1,-2,-3,\ldots\}$; and the prime numbers, the set of positive integers greater than 1 that are divisible only by 1 and themselves: $\{2,3,5,7,11,\ldots\}$.

## Rationals

One of the main subsets of the reals is the set of rational numbers, denoted $\mathbb{Q}$. This set is defined as all the numbers that can be expressed in the form a/b, where a and b are integers, $b \neq 0$. This form is called a fraction or ratio; a is known as the numerator, b the denominator.

e.g.     $\dfrac{-7}{5}, \dfrac{8}{6}, \dfrac{9}{-3}, \dfrac{5}{100}$

Note: the integers can all be expressed in the form a/b.

e.g.     $2 = \dfrac{2}{1}, \quad -3 = \dfrac{6}{-2}, \quad -5 = \dfrac{-5}{1}$

## Irrationals

The complement of the set of rationals is the irrationals, whose symbol is Q'. For now, they are defined as the set of real numbers that cannot be expressed in the form

$$\frac{a}{b}, \ b \neq 0.$$

6

# 2.2 ABSOLUTE VALUE

The absolute value of a real number A is defined as follows:

$$|A| = \begin{cases} A \text{ if } A \geq 0 \\ -A \text{ if } A < 0 \end{cases}$$

e.g.    $|5| = 5$,    $|-8| = -(-8) = 8$.

Absolute values follow the given rules:

A)  $|-A| = |A|$

B)  $|A| \geq 0$,  equality holding only if A=0

C)  $\left|\dfrac{A}{B}\right| = \dfrac{|A|}{|B|}$,    $B \neq 0$

D)  $|AB| = |A| \times |B|$

E)  $|A|^2 = A^2$

Absolute value can also be expressed on the real number line as the distance of the point represented by the real number from the point labeled 0.

3 unit lengths

So  $|-3| = 3$ because $-3$ is 3 units to the left of 0.


# 2.3 FUNDAMENTAL ALGEBRAIC LAWS

Note that $a, b, c \ \varepsilon \ \mathbb{R}$.

A)  Closure Law of Addition:

The sum of two real numbers is always a real number.

$$a + b = c.$$

B) Closure Law of Multiplication:

The product of two real numbers is always a real number.

$$a \times b = c.$$

C) Commutative Law of Addition:

$$a + b = b + a$$

Commutative refers to position. The sum of two real numbers is the same even if their positions are changed.

e.g.     $3 + 2 = 5 = 2 + 3.$

D) Commutative Law of Multiplication:

$$a \times b = b \times a$$

The product of two real numbers is the same even if their positions are changed.

e.g.     $3 \times 2 = 6 = 2 \times 3$

E) Associative Law of Addition:

$$(a + b) + c = a + (b + c)$$

Associative refers to grouping. The sum of any three real numbers is the same regardless of the way they are grouped.

e.g.     $(5 + 3) + 2 = 10 = 5 + (3 + 2).$

F) Associative Law of Multiplication:

$$(a \times b) \times c = a \times (b \times c)$$

The product of any three real numbers is the same, regardless of the way they are grouped.

e.g.     $(5 \times 3) \times 2 = 30 = 5 \times (3 \times 2)$

G) Additive Identity

There exists a real number 0 such that $a + 0 = a$. The number 0 is referred to as the additive identity.

H) Multiplicative Identity

There exists a real number 1 such that $a \times 1 = a$. The number 1 is referred to as the multiplicative identity.

I) Additive Inverse

For each real number a, there is a unique real number $-a$, called the additive inverse of a, such that $a + (-a) = 0$.

e.g.   $7 + (-7) = 0$.

J) Multiplicative Inverse

For every real number a, $a \neq 0$, there is a unique real number $\frac{1}{a}$, called the multiplicative inverse of a, such that $a \times \frac{1}{a} = 1$.

e.g.        $7 \times \frac{1}{7} = 1$

K) Zero Law

For every number a, $a \cdot 0 = 0$.

L) Distributive Law for Multiplication With Respect to Addition and Subtraction

$$a(b + c) = ab + ac$$
$$= ba + ca \quad \text{by the commutative law}$$
$$= (b + c)a$$

also        $$a(b - c) = ab - ac$$
$$= ba - ca$$
$$= (b - c)a$$

e.g.  1)  $3(4 + 5) = 3(4) + 3(5)$

$$= (4)3 + (5)3 = (4 + 5)3 = 27$$

2)  $3(5 - 4) = 3(5) - 3(4)$

$$= (5)3 - (4)3 = (5 - 4)3 = 3$$

These rules also hold for certain subsets of the reals $\mathbb{R}$, such as the rationals $\mathbb{Q}$. They do not hold for all subsets of $\mathbb{R}$, however; for instance, the integers $\mathbb{Z}$ do not contain multiplicative inverses for integers other than 1 or −1.

# 2.4 BASIC ALGEBRAIC OPERATIONS

A) To add two numbers with like signs, add their absolute values and prefix the sum with the common sign.

e.g.    $6 + 2 = 8$,    $(-6) + (-2) = -8$

B) To add two numbers with unlike signs, find the difference between their absolute values, and prefix the result with the sign of the number with the greater absolute value.

e.g.    $(-4) + 6 = 2$,    $15 + (-19) = -4$

C) To subtract a number b from another number a, change the sign of b and add to a.

Examples:    (1)    $10 - (3) = 10 + (-3) = 7$

(2)    $2 - (-6) = 2 + 6 = 8$

(3)    $(-5) - (-2) = -5 + (+2) = -3$

D) To multiply (or divide) two numbers having like signs, multiply (or divide) their absolute values and prefix the result with a positive sign.

Examples:    (1)  (5)(3) = 15

(2)  $\frac{-6}{-3}$ = 2

E)  To multiply (or divide) two numbers having unlike signs, multiply (or divide) their absolute values and prefix the result with a negative sign.

Examples:    (1)  (-2)(8) = -16

(2)  $\frac{9}{-3}$ = -3

# 2.5 OPERATIONS WITH FRACTIONS

To understand the operations on fractions, it is first desirable to understand what is known as factoring.

The product of two numbers is equal to a unique number. The two numbers are said to be factors of the unique number and the process of finding the two numbers is called factoring. It is important to note that when a number in a particular set is factored, then the factors of the number are also in the same set.

e.g.  The factors of 6 are

1)  1 and 6 since 1 × 6 = 6

2)  2 and 3 since 2 × 3 = 6.

A)  The value of a fraction remains unchanged, if its numerator and denominator are both multiplied or divided by the same number, other than zero.

e.g.    $\frac{1}{2} \times \frac{2}{2} = \frac{2}{4} = \frac{1}{2}$

This is because a fraction $\frac{b}{b}$ , b any number, is equal

to the multiplicative identity, 1.

11

B) To simplify a fraction is to convert it into a form in which numerator and denominator have no common factor other than 1.

e.g.
$$\frac{50}{25} = \frac{50 \div 25}{25 \div 25}$$

$$= \frac{2}{1} = 2$$

C) The algebraic sum of the fractions having a common denominator is a fraction whose numerator is the algebraic sum of the numerators of the given fractions and whose denominator is the common denominator.

e.g.
$$\frac{11}{3} + \frac{5}{3} = \frac{11 + 5}{3} = \frac{16}{3}$$

Similarly, for subtraction,

$$\frac{11}{3} - \frac{5}{3} = \frac{11 - 5}{3} = \frac{6}{3} = 2$$

D) To find the sum of two fractions having different denominators, it is necessary to find the lowest common denominator, (LCD), of the different denominators and convert the fractions into equivalent fractions having the lowest common denominator as a denominator.

e.g.
$$\frac{11}{6} + \frac{5}{16} = ?$$

To find the LCD, we must first find the prime factors of the two denominators.

$$6 = 2 \cdot 3$$

$$16 = 2 \cdot 2 \cdot 2 \cdot 2$$

$$LCD = 2 \cdot 2 \cdot 2 \cdot 2 \cdot 3 = 48$$

Note we do not need to repeat the 2 that appears in both the factors of 6 and 16.

We now rewrite 11/6, 5/16 to have 48 as their denominator.

$$\frac{11}{6} \cdot \frac{8}{8} = \frac{88}{48}$$

$$\frac{5}{16} \cdot \frac{3}{3} = \frac{15}{48}$$

We may now apply rule 3 to find $\frac{11}{6} + \frac{5}{16} = \frac{103}{48}$.

E) The product of two or more fractions produces a fraction whose numerator is the product of the numerators of the given fractions and whose denominator is the product of the denominators of the given fractions.

e.g. $\quad \frac{2}{3} \cdot \frac{1}{5} \cdot \frac{4}{7} = \frac{8}{105}$

F) The quotient of two given fractions is obtained by inverting the divisor and then multiplying.

e.g. $\quad \frac{8}{9} \div \frac{1}{3} = \frac{8}{9} \times \frac{3}{1} = \frac{8}{3}$

# 2.6 DECIMALS

If we divide the denominator of a fraction into its numerator, we obtain a decimal form for it. This form attaches significance to the placement of an integer relative to a decimal point. The first place to the left of the decimal point is the units place; the second to the left, is the tens; third, the hundreds, etc. The first place to the right of the decimal point is the tenths, the second the hundredths, etc. The integer in each place tells how many of the values of that place the given number has.

> Example: 721 has seven hundreds, two tens, and one unit. .584 has five tenths, eight hundredths, and four thousandths.

Since a rational number is of the form $\frac{a}{b}$, $b \neq 0$, then all rational numbers can be expressed as decimals by dividing b into a. The resulting decimal is either a terminating decimal, meaning that b divides a with remainder 0 after a certain point; or repeating, meaning that b continues to divide a so that the decimal has a repeating pattern of integers.

e.g.  A) $\frac{1}{2}$ = .5

B) $\frac{1}{3}$ = .333...

C) $\frac{11}{16}$ = .6875

D) $\frac{2}{7}$ = .285714285714....

A) and C) are terminating decimals; B) and D) are repeating decimals. This explanation allows us to define irrational numbers as numbers whose decimal form is nonterminating and nonrepeating.

e.g.  $\sqrt{2}$ = 1.414...

$\sqrt{3}$ = 1.732....

So the set of reals is the union of the set of rationals and the set of irrationals ($\mathbb{R} = \mathbb{Q} \cup \mathbb{Q}'$).

# 2.7 IMAGINARY AND COMPLEX NUMBERS

Imaginary numbers

If a number a is multiplied by itself to produce a new number b, b is called the square of a and a the square root of b, denoted $a = \sqrt{b}$.

e.g.  $3 \cdot 3 = 9$,  $\sqrt{9} = 3$

$3^2$, read "3 to the second power" or "3 squared", indicates that 3 is to be used as a factor of the expression $3^2$, i.e. $3^2 = 3 \cdot 3$.

According to the law of signs for real numbers, the square of a positive or negative number is always positive. This means that it is impossible to take the square root of a negative number in the real number system. In order to make this possible, the symbol i is defined as $i \equiv \sqrt{-1}$, $i^2 = -1$. i is called an imaginary number, as is any multiple of i by a real number. This set is denoted $\mathbb{R}'$.

## Complex Numbers ($\mathbb{C}$)

A complex number is a combination of real and imaginary numbers of the form a + bi where a and b are real and i is defined as above. a is called the real part while b is called the imaginary part of a + bi.

Both the real and the imaginary number set are subsets of $\mathbb{C}$. $\mathbb{R} = \{ a + bi \mid b = 0 \}$ and $\mathbb{R}' = \{ a + bi \mid a = 0 \}$. Figure 2.3 uses a Venn diagram to illustrate the relationships of the various number systems to each other, while Figure 2.4 uses the tree form.

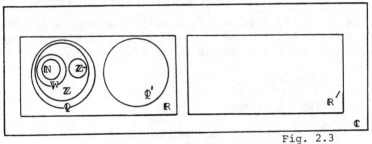

Fig. 2.3

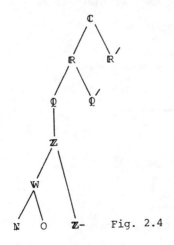

Fig. 2.4

15

# CHAPTER 3

# EXPONENTS AND RADICALS

## 3.1 EXPONENTS

Given the expression $a^n = b$, where a, n, and $b \in \mathbf{R}$, a is called the base, n is called the exponent or power.

e.g. In $3^2$, 3 is the base, 2 is the exponent. If n is a positive integer and if x and y are real numbers such that $x^n = y$, then x is said to be an nth root of y, written

$$x = \sqrt[n]{y} = y^{1/n}.$$

Positive Integral Exponent:

If n is a positive integer, then $a^n$ represents the product of n factors each of which is a.

Negative Integral Exponent:

If n is a positive integer,

$$a^{-n} = \frac{1}{a^n} \qquad a \neq 0$$

e.g. $2^{-4} = \frac{1}{2^4} = \frac{1}{16}$

Positive Fractional Exponent:

$$a^{m/n} = \sqrt[n]{a^m}$$

where m and n are positive integers.

e.g. $\quad 4^{3/2} = \sqrt[2]{4^3} = \sqrt{64} = 8$

Negative Fractional Exponent:

$$a^{-m/n} = \frac{1}{a^{m/n}}$$

e.g. $\quad 27^{-2/3} = \dfrac{1}{27^{2/3}} = \dfrac{1}{\sqrt[3]{27^2}} = \dfrac{1}{\sqrt[3]{729}} = \dfrac{1}{9}$

Zero Exponent:

$$a^0 = 1 \ , \ a \neq 0$$

General Laws of Exponents:

A) $\quad a^p a^q = a^{p+q}$

B) $\quad (a^p)^q = a^{pq}$

C) $\quad \dfrac{a^p}{a^q} = a^{p-q} \ , \quad a \neq 0$

D) $\quad (ab)^p = a^p b^p$

E) $\quad \left(\dfrac{a}{b}\right)^p = \dfrac{a^p}{b^p} \ , \quad b \neq 0$

# 3.2 RADICALS

A radical is an expression of the form $\sqrt[n]{a}$ which denotes the nth root of a positive integer a; n is the index of the radical and the number a is the radicand. The index is usually omitted if n = 2.

Laws for radicals are the same as laws for exponents, since

$$\sqrt[n]{a} = a^{\frac{1}{n}}, \qquad n \neq 0$$

Some of these laws are:

A) $(\sqrt[n]{a})^n = a$

B) $\sqrt[n]{ab} = \sqrt[n]{a}\ \sqrt[n]{b}$

C) $\sqrt[n]{a/b} = \sqrt[n]{a}/\sqrt[n]{b}$ $\qquad b \neq 0$

D) $\sqrt[n]{a^m} = (\sqrt[n]{a})^m$

E) $\sqrt[m]{\sqrt[n]{a}} = \sqrt[mn]{a}$

A radical is said to be in simplest form if:

A) All perfect nth powers have been removed from the radical.

e.g. $\sqrt[3]{8x^5} = \sqrt[3]{(2x)^3 \cdot x^2} = 2x(\sqrt[3]{x^2})$.

B) The index of the radical is as small as possible.

C) There aren't any fractions present in the radicand.

Two radicals are said to be similar if they have the same index and the same radicand.

To algebraically add or subtract two or more radicals, reduce each given radical to the simplest form, and add or subtract terms with the same radicals.

e.g. $\sqrt{27} + \sqrt{12} = \sqrt{3^3 \cdot 3} + \sqrt{2^2 \cdot 3} = 3\sqrt{3} + 2\sqrt{3} = 5\sqrt{3}$

To multiply two or more radicals with the same radicands, write the radicals in the form $a^x$, then apply the law

$$a^x a^y = a^{x+y}$$

e.g. $\sqrt{27} + \sqrt{12} = \sqrt{3^2 \cdot 3} + \sqrt{2^2 \cdot 3} = 3\sqrt{3} + 2\sqrt{3} = 5\sqrt{3}$

$$= 2^{(\frac{1}{2} + \frac{1}{5} + \frac{4}{3})} = 2^{\frac{61}{30}} = \sqrt[30]{2^{61}}$$

To divide two radicals with the same radicands, write the radicals in the form $a^x$, then apply the law

$$\frac{a^x}{a^y} = a^{x-y}$$

For example:

$$(\sqrt{5}) \div (\sqrt[3]{5}) = (5^{\frac{1}{2}}) \div (5^{\frac{1}{3}})$$

$$= 5^{(\frac{1}{2} - \frac{1}{3})} = 5^{\frac{1}{6}}$$

$$= \sqrt[6]{5}$$

# 3.3 SCIENTIFIC NOTATION

A real number expressed in scientific notation is written as a product of a real number n and an integral power of 10; the value of n is $1 \leq n < 10$.

e.g.

| | Number | Scientific Notation |
|---|---|---|
| 1) | 1956. | $1.956 \times 10^3$ |
| 2) | .0036 | $3.6 \times 10^{-3}$ |
| 3) | 59600000. | $5.96 \times 10^7$ |

# CHAPTER 4

# POLYNOMIALS AND RATIONAL EXPRESSIONS

## 4.1 TERMS AND EXPRESSIONS

A <u>variable</u> is defined as a placeholder, which can take on any of several values at a given time; it is usually represented by one of the last letters of the alphabet such as x, y, and z. A <u>constant</u> is a symbol which takes on only one value at a given time. If the value of the constant is unknown, it is usually denoted by the letters a, b, or c. $\pi$, 5, 8/17, -i are constants.

A <u>term</u> is a constant, a variable, a product of constants and variables, a quotient of constants and variables, or a combination of products and quotients. For example: 7.76, 3x, xyz, x/5, 5z/x, $(0.99)x^2$ are terms. If a term is a combination of constants and variables, the constant part of the term is referred to as the <u>coefficient</u> of the variable. If a variable is written without a coefficient, the coefficient is assumed to be 1.

e.g.
$3x^2$
coefficient: 3
variable: x

$y^3 = (1)y^3$
coefficient: 1
variable: y

An <u>expression</u> is a collection of one or more terms. If the number of terms is greater than 1, the expression is said to be the sum of terms.

e.g.   9, 9xy, 6x + x/3, 8yz - 2x.

20

# 4.2 THE POLYNOMIAL

An algebraic expression consisting of only one term is called a <u>monomial</u>. An algebraic expression consisting of two terms is called a <u>binomial</u>. An algebraic expression consisting of three terms is called a <u>trinomial</u>. In general, an algebraic expression consisting of two or more terms is called a <u>multinomial</u> or polynomial.

A <u>polynomial in x</u>, denoted P(x), consists of one or more terms such that the terms are either an integral constant or the product of an integral constant and a positive integral power of x.

e.g.   $5x^3 + 2x^2 + 3$ is a polynomial in x.

$2x^2 + x^{\frac{1}{2}} - 1$  is not a polynomial in x.

$9x^3 + 3x^{-2} + 4$ is not a polynomial in x.

The degree of a monomial is the sum of the exponents of the variables. The degree of a monomial with no variables is 0.

e.g.   $5x^2$ has degree 2

$3x^3y^2z$ has degree 6

9 has degree 0.

The degree of a polynomial is equal to the exponent of that term with the highest power of x whose coefficient is not 0.

e.g.   $5x^4 + 7x + 12$ has degree 4.

# 4.3 ALGEBRAIC OPERATIONS WITH POLYNOMIALS

Addition of polynomials is achieved by combining like

terms, defined as terms which differ only in numerical coefficients.

e.g. $(x^2 - 3x + 5) + (4x^2 + 6x - 3)$

Note: Parenthesis are used to distinguish polynomials.

By using the commutative and associative laws, we can rewrite $P(x)$ as:

$P(x) = (x^2 + 4x^2) + (6x - 3x) + (5 - 3)$

Using the distributive law yields

$(1 + 4)x^2 + (6 - 3)x + (5 - 3)$

$= 5x^2 + 3x + 2.$

Subtraction of two polynomials is achieved by first changing the sign of all terms in the expression which is being subtracted and then adding this result to the other expression.

e. g. $(5x^2 + 4y^2 + 3z^2) - (4xy + 7y^2 - 3z^2 + 1)$

$= 5x^2 + 4y^2 + 3z^2 - 4xy - 7y^2 + 3z^2 - 1$

$= (5x^2) + (4y^2 - 7y^2) + (3z^2 + 3z^2) - 4xy - 1$

$= (5x^2) + (-3y^2) + (6z^2) - 4xy - 1.$

Multiplication of two or more monomials is achieved by using the laws of exponents, the rules of signs, and the commutative and associative laws of multiplication.

e.g. $(y^2)(5)(6y^2)(yz)(2z^2)$

$= (1)(5)(6)(1)(2)(y^2)(y^2)(yz)(z^2)$

$= (60)[(y^2)(y^2)(y)][(z)(z^2)]$

$= 60(y^5)(z^3)$

$= 60y^5 z^3.$

Multiplication of a polynomial by a monomial is achieved by multiplying each term of the polynomial by the monomials and combining the results.

e.g. $(4x^2 + 3y) \times (6xz^2)$

$= 24x^3z^2 + 18xyz^2.$

Multiplication of a polynomial by a polynomial is achieved by multiplying each of the terms of one polynomial by each of the terms of the other polynomial and combining the result.

e.g. $(5y + z + 1) \times (y^2 + 2y)$

$= [(5y) \times (y^2) + (5y) \times (2y)]$

$+ [(z) \times (y^2) + (z) \times (2y)]$

$+ [(1) \times (y^2) + (1) \times (2y)]$

$= (5y^3 + 10y^2) + (y^2z + 2yz) + (y^2 + 2y)$

$= (5y^3) + (10y^2 + y^2) + (y^2z) + (2yz) + (2y)$

$= 5y^3 + 11y^2 + y^2z + 2yz + 2y.$

Division of a monomial by a monomial is achieved by finding the quotient of the constant coefficients and the quotients of the variable factors, followed by the multiplication of these quotients.

e.g. $6xyz^2 \div 2y^2z$

$= (6/2)(x/1)(y/y^2)(z^2/z)$

$= 3xy^{-1}z$

$= 3xz/y$

Division of a polynomial by a polynomial is achieved by following the given procedure called Long Division.

Step 1: The terms of both the polynomials are arranged in order of ascending or descending powers of one variable.

Step 2: The first term of the dividend is divided by the first term of the divisor which gives the first term of the quotient.

Step 3: The divisor is multiplied by the first term of the quotient and the result is subtracted from the dividend.

Step 4: Using the remainder obtained from step 3 as the new dividend, steps 2 and 3 are repeated until the remainder is zero or the degree of the remainder is less than the degree of the divisor.

Step 5: The result is written as follows:

$$\frac{\text{dividend}}{\text{divisor}} = \text{quotient} + \frac{\text{remainder}}{\text{divisor}},$$

$$\text{divisor} \neq 0$$

e.g. $(2x^2 + x + 6) \div (x + 1)$

$$
\begin{array}{r}
2x - 1 \\
x + 1 \overline{\smash{)}\, 2x^2 + x + 6} \\
- \underline{(2x^2 + 2x)} \\
-x + 6 \\
- \underline{(-x - 1)} \\
7
\end{array}
$$

The result is

$$(2x^2 + x + 6) \div (x + 1) = 2x - 1 + \frac{7}{x + 1}$$

# 4.4 POLYNOMIAL FACTORIZATION

To factor a polynomial completely is to find the prime factors of the polynomial with respect to a specified set of numbers, i.e. to express it as a product of polynomials whose coefficients are members of that set.

The following concepts are important while factoring polynomials.

The factors of an algebraic expression consist of two or more algebraic expressions which when multiplied together produce the given algebraic expression.

A prime factor is a polynomial with no factors other than itself and 1. The least common multiple for a set of numbers is the smallest quantity divisible by every number of the set. For algebraic expressions the least common multiple is the polynomial of lowest degree and smallest numerical coefficients for which each of the given expressions will be a factor.

The greatest common factor for a set of numbers is the largest factor that is common to all members of the set.

For algebraic expressions the greatest common factor is the polynomial of highest degree and largest numerical coefficients which is a factor of all the given expressions.

Some important formulae, useful for the factoring of polynomials, are listed below.

$a(c + d) = ac + ad$

$(a + b)(a - b) = a^2 - b^2$

$(a + b)(a + b) = (a + b)^2 = a^2 + 2ab + b^2$

$(a - b)(a - b) = (a - b)^2 = a^2 - 2ab + b^2$

$(x + a)(x + b) = x^2 + (a + b)x + ab$

$(ax + b)(cx + d) = acx^2 + (ad + bc)x + bd$

$(a + b)(c + d) = ac + bc + ad + bd$

$(a + b)(a + b)(a + b) = (a + b)^3 = a^3 + 3a^2b + 3ab^2 + b^3$

$(a - b)(a - b)(a - b) = (a - b)^3 = a^3 - 3a^2b + 3ab^2 - b^3$

$(a - b)(a^2 + ab + b^2) = a^3 - b^3$

$(a + b)(a^2 - ab + b^2) = a^3 + b^3$

$(a + b + c)^2 = a^2 + b^2 + c^2 + 2ab + 2ac + 2bc$

$(a - b)(a^3 + a^2b + ab^2 + b^3) = a^4 - b^4$

$(a - b)(a^4 + a^3b + a^2b^2 + ab^3 + b^4) = a^5 - b^5$

$(a - b)(a^5 + a^4b + a^3b^2 + a^2b^3 + ab^4 + b^5) = a^6 - b^6$

$(a - b)(a^{n-1} + a^{n-2}b + a^{n-3}b^2 + \ldots + ab^{n-2} + b^{n-1})$

$$= a^n - b^n$$

where n is any positive integer $(1,2,3,4,\ldots)$.

$$(a + b)(a^{n-1} - a^{n-2}b + a^{n-3}b^2 - \ldots - ab^{n-2} + b^{n-1})$$

$$= a^n + b^n$$

where n is any positive odd integer $(1,3,5,7,\ldots)$.

The procedure for factoring a polynomial completely is as follows.

Step 1: First find the greatest common factor if there is any. Then examine each factor remaining for greatest common factors.

Step 2: Continue factoring the factors obtained in step 1 until all factors other than monomial factors are prime.

e.g. Factoring $4 - 16x^2$,

$$4 - 16x^2 = 4(1 - 4x^2) = 4(1 + 2x)(1 - 2x)$$

# 4.5 OPERATIONS WITH FRACTIONS AND RATIONAL EXPRESSIONS

A _rational_ expression is an algebraic expression which can be written as the quotient of two polynomials, $\frac{A}{B}$, $B \neq 0$.

e.g. $\quad \frac{9}{4}, \quad \frac{3x^2 + 5x}{y + 3}, \quad \frac{9y}{10z}$

To reduce a given fraction or a rational expression to its simplest form is to reduce the fraction or expression into an equivalent form such that its numerator and denominator have no common factor other than 1.

The operations performed on fractions are in a similar manner applicable to rational expressions.

A fraction that contains one or more fractions in either its numerator or denominator – or in both its numerator and denominator – is called a complex fraction.

e.g.
$$\frac{\frac{1}{x}}{\frac{5}{y}}, \quad \frac{\frac{1}{2}}{3}, \quad \frac{1 + \frac{y}{x}}{1 - \frac{4}{x^2 + 1}}$$

The procedure for simplifying complex fractions is as follows:

First, the terms in the numerator and denominator are separately combined. Then the combined term of the numerator is divided by the combined term of the denominator to obtain a simplified fraction.

e.g.
$$\frac{1 - \frac{5}{x} + \frac{6}{x^2}}{1 - \frac{6}{x} + \frac{8}{x^2}}$$

Combining the numerator we get

$$1 - \frac{5}{x} + \frac{6}{x^2} = \frac{x^2 - 5x + 6}{x^2}$$

$$= \frac{(x - 3)(x - 2)}{x^2}$$

Combining the denominator we get

$$1 - \frac{6}{x} + \frac{8}{x^2} = \frac{x^2 - 6x + 8}{x^2}$$

$$= \frac{(x - 4)(x - 2)}{x^2}$$

Dividing resultant numerator by resultant denominator we get:

$$\frac{\frac{(x - 3)(x - 2)}{x^2}}{\frac{(x - 4)(x - 2)}{x^2}} = \frac{(x - 3)(x - 2)}{x^2} \times \frac{x^2}{(x - 4)(x - 2)}$$

$$= \frac{x - 3}{x - 4}$$

# CHAPTER 5

# EQUATIONS

## 5.1 EQUATIONS

An equation is defined as a statement of equality of two separate expressions known as members.

A conditional equation is an equation which is true for only Cartesian values of the unknowns (variables) invoked.

e.g. $y + 6 = 11$ is true for $y = 5$.

An equation which is true for all permissible values of the unknown in question is called an identity. For example, $2x = \frac{4}{2}x$ is an identity of $x \in \mathbb{R}$, i.e., it is true for all reals.

The values of the variables that satisfy a conditional equation are called solutions of the conditional equation; the set of all such values is known as the solution set.

The solution to an equation $f(x) = 0$ is called the root of the equation.

Equations with the same solutions are said to be equivalent equations.

A statement of equality between two expressions containing rational coefficients and whose exponents are integers is called a rational integral equation. The degree of the equation is given by the term with highest power, as shown below:

28

$$a_nx^n + a_{n-1}x^{n-1} + a_{n-2}x^{n-2} + \ldots + a_1x + a_0 = 0$$

where $a_n \neq 0$, the $a_i$, $i = 1\ldots n$ are rational constant coefficients and n is a positive integer.

# 5.2 BASIC LAWS OF EQUALITY

A) Replacing an expression of an equation by an equivalent expression results in an equation equivalent to the original one.

E.g.  Given the equation below

$$3x + y + x + 2y = 15$$

We know that for the left side of this equation we can apply the commutative and distributive laws to get:

$$3x + y + x + 2y = 4x + 3y.$$

Since these are equivalent, we can replace the expression in the original equation with the simpler form to get:

$$4x + 3y = 15.$$

B) The addition or subtraction of the same expression on both sides of an equation results in an equivalent equation to the original one:

E.g.  Given the equation

$$y + 6 = 10,$$

we can add (-6) to both sides

$$y + 6 + (-6) = 10 + (-6)$$

to get   $y + 0 = 10 - 6 \Rightarrow y = 4.$

So $y + 6 = 10$ is equivalent to $y = 4$.

C) The multiplication or division on both sides of an equation by the same expression results in an equivalent equation to the original.

E.g. $3x = 6 \Rightarrow \dfrac{3x}{3} = \dfrac{6}{3} \Rightarrow x = 2$.

$3x = 6$ is equivalent to $x = 2$.

D) If both members of an equation are raised to the same power, then the resultant equation is equivalent to the original equation.

Example: $a = x^2 y$, $(a)^2 = (x^2 y)^2$, and $a^2 = x^4 y^2$.

This applies for negative and fractional powers as well.

E.g. $x^2 = 3y^4$.

If we raise both members to the $-2$ power we get

$$(x^2)^{-2} = (3y^4)^{-2}$$
$$\frac{1}{(x^2)^2} = \frac{1}{(3y^4)^2}$$
$$\frac{1}{x^4} = \frac{1}{9y^8}$$

If we raise both members to the $\frac{1}{2}$ power which is the same as taking the square root, we get:

$$(x^2)^{\frac{1}{2}} = (3y^4)^{\frac{1}{2}}$$
$$x = \sqrt{3}\ y^2$$

E) The reciprocal of both members of an equation are equivalent to the original equation.

Note: The reciprocal of zero is undefined.

$$\frac{2x + y}{z} = \frac{5}{2} \qquad \frac{z}{2x + y} = \frac{2}{5}$$

# 5.3 EQUATIONS WITH ABSOLUTE VALUES

When evaluating equations containing absolute values,

proceed as follows:   Example:   $|5 - 3x| = 7$   is   valid   if either

$$5 - 3x = 7 \quad \text{or} \quad 5 - 3x = -7$$

$$-3x = 2 \qquad\qquad -3x = -12$$

$$x = \frac{-2}{3} \qquad\qquad x = 4$$

The solution set is therefore $x = (-2/3, 4)$.

# CHAPTER 6

# LINEAR EQUATIONS AND SYSTEMS OF LINEAR EQUATIONS

## 6.1 LINEAR EQUATIONS

A linear equation in one variable is one that can be put into the form $ax + b = 0$, where a and b are constants, $a \neq 0$.

To solve a linear equation means to transform it in the form $x = \frac{-b}{a}$.

A) If the equation has unknowns on both sides of the equality, it is convenient to put similar terms on the same side.

E.g. $4x + 3 = 2x + 9$

$4x + 3 - 2x = 2x + 9 - 2x$

$(4x - 2x) + 3 = (2x - 2x) + 9$

$2x + 3 = 0 + 9$

$2x + 3 - 3 = 0 + 9 - 3$

$2x = 6$

$\frac{2x}{2} = \frac{6}{2}$

$$x = 3.$$

B) If the equation appears in fractional form, it is necessary to transform it, using cross-multiplication, and then repeating the same procedure as in A), we obtain:

$$\frac{3x + 4}{3} \times \frac{7x + 2}{5}$$

By using cross-multiplication we would obtain:

$$3(7x + 2) = 5(3x + 4).$$

This is equivalent to:

$$21x + 6 = 15x + 20,$$

which can be solved as in A):

$$21x + 6 = 15x + 20$$

$$21x - 15x + 6 = 15x - 15x + 20$$

$$6x + 6 - 6 = 20 - 6$$

$$6x = 14$$

$$x = \frac{14}{6}$$

$$x = \frac{7}{3}$$

C) If there are radicals in the equation, it is necessary to square both sides and then apply A)

$$\sqrt{3x + 1} = 5$$

$$(\sqrt{3x + 1})^2 = 5^2$$

$$3x + 1 = 25$$

$$3x + 1 - 1 = 25 - 1$$

$$3x = 24$$

$$x = \frac{24}{3}$$

$$x = 8$$

# 6.2 LINEAR EQUATIONS
## IN TWO VARIABLES

Equations of the form ax + by = c, where a, b, c are constants and a,b $\neq$ 0 are called linear equations in two variables.

The solution set for a linear equation in two variables is the set of all x and y values for which the equation is true. An element in the solution set is called an ordered pair (x,y) where x and y are two of the values that together satisfy the equation. The x value is always first and is called x-coordinate. The y value is always second and is called the y-coordinate.

# 6.3 GRAPHING THE SOLUTION SET

The solution set of the equation ax + by = c can be represented by graphing the ordered pairs that satisfy the equation on a rectangular coordinate system. This is a system where two real number lines are drawn at right angles to each other. The point where the two lines intercept is called the origin and is associated to the ordered pair (0,0).

To plot a certain ordered pair (x,y) move x units along the x-axis in the direction indicated by the sign of x, then move y units along the y-axis in the direction indicated by the sign of y.

Note that movement to the right or up is positive, while movement to the left or down is negative.

Example: Graph the following points: (1,2), (-3,2), (-2,-1), (1,-1).

To graph a linear equation in two variables it is necessary to graph its solution set, that is, draw a line through the points whose coordinates satisfy the equation.

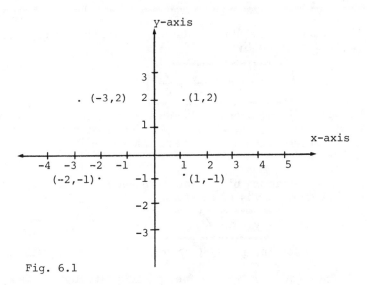

Fig. 6.1

The resultant graph of a linear equation in two variables is a straight line.

There are several ways of graphing this line, two of them are shown below:

A) Plot 2 or more ordered pairs that satisfy the equation and then connect them.

B) Plot the points $A\left(\dfrac{c}{a},\ 0\right)$ and $B\left(0,\ \dfrac{c}{b}\right)$ that correspond to the points where the line intercepts the x-axis and y-axis respectively, as shown:

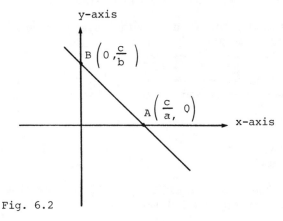

Fig. 6.2

The slope of the line containing two points $(x_1, y_1)$ and $(x_2, y_2)$ is given by:

$$\text{Slope} = m = \frac{y_2 - y_1}{x_2 - x_1}.$$

Horizontal lines have a slope of zero and the slope of vertical lines is undefined. Parallel lines have equal slopes and perpendicular lines have slopes which are negative reciprocals.

The equation of a line with slope m passing through a point $Q(x_0, y_0)$ is of the form:

$$y - y_0 = m(x - x_0).$$

This is called the point-slope form of a linear equation.

The equation of a line passing through $Q(x_1, y_1)$ and $P(x_2, y_2)$ is given by

$$\frac{x - x_1}{x_1 - x_2} = \frac{y - y_1}{y_1 - y_2}$$

This is the two-point form of a linear equation.

The equation of a line intersecting the x-axis at $(x_0, 0)$ and the y-axis at $(0, y_0)$ is given by:

$$\frac{x}{x_0} + \frac{y}{y_0} = 1.$$

This is the intercept form of a linear equation.

The equation of a line with slope m intersecting the y-axis at $(0, b)$ is given by:

$$y = mx + b.$$

This is the slope-intercept form of a linear equation.

Problems on Linear Equations:

A) Find the slope, the y-intercept, and the x-intercept of the equation $2x - 3y - 18 = 0$.

Solution: The equation $2x - 3y - 18 = 0$ can be written in the form of the general linear equation, $ax + by = c$.

$$2x - 3y - 18 = 0$$

$$2x - 3y = 18$$

To find the slope and y-intercept we derive them from the formula of the general linear equation ax + by = c. Dividing by b and solving for y we obtain:

$$\frac{a}{b}x + y = \frac{c}{b}$$

$$y = \frac{c}{b} - \frac{a}{b}x$$

where $\frac{-a}{b}$ = slope and $\frac{c}{b}$ = y-intercept.

To find the x-intercept, solve for x and let y = 0:

$$x = \frac{c}{b} - \frac{b}{a}y$$

$$x = \frac{c}{a}$$

In this form we have a = 2, b = -3, and c = 18. Thus,

$$\text{slope} = -\frac{a}{b} = -\frac{2}{-3} = \frac{2}{3}$$

$$\text{y-intercept} = \frac{c}{b} = \frac{18}{-3} = -6$$

$$\text{x-intercept} = \frac{c}{a} = \frac{18}{2} = 9$$

B) Find the equation for the line passing through (3,5) and (-1,2).

Solution A:   We use the two-point form with $(x_1, y_1) = (3,5)$ and $(x_2, y_2) = (-1,2)$. Then

$$\frac{y - y_1}{x - x_1} = m = \frac{y_2 - y_1}{x_2 - x_1}$$

$$\frac{y_2 - y_1}{x_2 - x_1} = \frac{2 - 5}{-1 - 3} \quad \text{thus} \quad \frac{y - 5}{x - 3} = \frac{-3}{-4}$$

Cross multiply, $-4(y - 5) = -3(x - 3)$.

Distributing, $-4y + 20 = -3x + 9$

$$3x - 4y = -11.$$

Solution B: Does the same equation result if we let $(x_1, y_1) = (-1,2)$ and $(x_2, y_2) = (3,5)$?

$$\frac{y_2 - y_1}{x_2 - x_1} = \frac{5 - 2}{3 - (-1)} \quad \text{thus} \quad \frac{y - 2}{x + 1} = \frac{3}{4}.$$

Cross multiply, $4(y - 2) = 3(x + 1)$

$$3x - 4y = -11.$$

Hence, either replacement results in the same equation.

C) (a) Find the equation of the line passing through $(2,5)$ with slope 3.

(b) Suppose a line passes through the y-axis at $(0,b)$. How can we write the equation if the point-slope form is used?

Solution: (a) In the point-slope form, let $x_1 = 2$, $y_1 = 5$, and $m = 3$. The point-slope form of a line is:

$$y - y_1 = m(x - x_1)$$

$$y - 5 = 3(x - 2)$$

$$y - 5 = 3x - 6 \quad \text{Distributive property}$$

$$y = 3x - 1 \quad \text{Transposition}$$

(b) $\quad y - b = m(x - 0)$

$$y = mx + b.$$

D) Graph the function defined by $3x - 4y = 12$.

Solution: Solve for y:

$$3x - 4y = 12$$

$$-4y = 12 - 3x$$

$$y = -3 + \frac{3}{4}x$$

$$y = \frac{3}{4}x - 3.$$

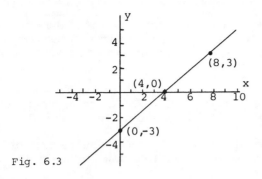

Fig. 6.3

The graph of this function is a straight line since it is of the form $y = mx + b$. The y-intercept is the point $(0,-3)$ since for $x = 0$, $y = b = -3$. The x-intercept is the point $(4,0)$ since for $y = 0$, $x = (y + 3) \cdot \frac{4}{3} = (0 + 3) \cdot \frac{4}{3} = 4$. These two points. $(0,-3)$ and $(4,0)$ are sufficient to determine the graph (see the figure). A third point, $(8,3)$, satisfying the equation of the function is plotted as a partial check of the intercepts. Note that the slope of the line is $m = \frac{3}{4}$. This means that y increases 3 units as x increases 4 units anywhere along the line.

# 6.4 SYSTEMS OF LINEAR EQUATIONS

A system of linear equations is a set of one or more linear equation as shown below:

$$\begin{cases} 2x + 4y = 11 \\ -5x + 3y = 5 \end{cases}$$

The set shown above is a system of linear equations with two variables, or unknowns.

There are several ways to solve systems of linear equations in two variables:

Method 1: Addition or subtraction - if necessary multiply the equations by numbers that will make the coefficients of one unknown in the resulting equations numerically equal. If the signs of equal coefficients are the same, subtract the equation, otherwise add.

The result is one equation with one unknown; we solve it and substitute the value into the other equations to find the unknown that we first eliminated.

Method 2: Substitution - find the value of one unknown in terms of the other, substitute this value in the other equation and solve.

Method 3: Graph - graph both equations. The point of intersection of the drawn lines is a simultaneous solution for the equations and its coordinates correspond to the answer that would be found analytically.

If the lines are parallel they have no simultaneous solution.

Dependent equations are equations that represent the same line, therefore every point on the line of a dependent equation represents a solution. Since there is an infinite number of points there is an infinite number of simultaneous solutions, for example:

$$\begin{cases} 2x + y = 8 \\ 4x + 2y = 16 \end{cases}$$

The equations above are dependent, they represent the same line, all points that satisfy either of the equations are solutions of the system.

A system of linear equations is consistent if there is only one solution for the system.

A system of linear equations is inconsistent if it does not have any solutions.

Example of a consistent system:
Find the point of intersection of the graphs of the equations:

$$\begin{cases} x + y = 3, \\ 3x - 2y = 14. \end{cases}$$

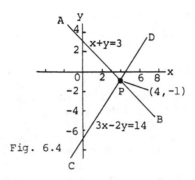

Fig. 6.4

Solution: To solve these linear equations, solve for y in terms of x. The equations will be in the form y = mx + b, where m is the slope and b is the intercept on the y-axis.

$$x + y = 3$$

$$y = 3 - x \quad \text{subtract x from both sides}$$

$$3x - 2y = 14 \quad \text{subtract 3x from both sides}$$

$$-2y = 14 - 3x \quad \text{divide by -2.}$$

$$y = -7 + \frac{3}{2}x$$

The graphs of the linear functions, y = 3 - x and y = -7 + $\frac{3}{2}$x, can be determined by plotting only two points. For example, for y = 3 - x, let x = 0, then y = 3. Let x = 1, then y = 2. The two points on this first line are (0,3) and (1,2). For y = -7 + $\frac{3}{2}$x, let x = 0, then y = -7. Let x = 1, then y = -5½. The two points on this second line are (0,-7) and (1,-5½).

To find the point of intersection P of

$$x + y = 3$$

and

$$3x - 2y = 14,$$

solve them algebraically. Multiply the first equation by 2. Add these two equations to eliminate the variable y.

41

$$2x + 2y = 6$$
$$3x - 2y = 14$$
$$\overline{5x \phantom{+ 2y} = 20}$$

Solve for x to obtain x = 4.  Substitute this into y = 3 – x to get y = 3 – 4 = –1.  P is (4,–1).  AB is the graph of the first equation, and CD is the graph of the second equation.  The point of intersection P of the two graphs is the only point on both lines.  The coordinates of P satisfy both equations and represent the desired solution of the problem.  From the graph, P seems to be the point (4,–1). These coordinates satisfy both equations, and hence are the exact coordinates of the point of intersection of the two lines.

To show that (4,–1) satisfies both equations, substitute this point into both equations.

| | |
|---|---|
| x + y = 3 | 3x – 2y = 14 |
| 4 + (–1) = 3 | 3(4) – 2(–1) = 14 |
| 4 – 1 = 3 | 12 + 2 = 14 |
| 3 = 3 | 14 = 14 |

Example of an inconsistent system.

Solve the equations 2x + 3y = 6 and 4x + 6y = 7 simultaneously.

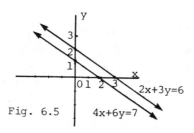

Fig. 6.5

Solution:  We have 2 equations in 2 unknowns,

$$2x + 3y = 6 \qquad\qquad (1)$$

and

$$4x + 6y = 7 \qquad\qquad (2)$$

There are several methods to solve this problem. We have chosen to muliply each equation by a different number so that when the two equations are added, one of the variables drops out.  Thus

multiplying equation (1) by 2:  $4x + 6y = 12$  (3)

multiplying equation (2) by -1:  $\underline{-4x - 6y = -7}$  (4)

adding equations (3) and (4):  $0 = 5$

We obtain a peculiar result!

Actually, what we have shown in this case is that if there were a simultaneous solution to the given equations, then 0 would equal 5.  But the conclusion is impossible; therefore there can be no simultaneous solution to these two equations, hence no point satisfying both.

The straight lines which are the graphs of these equations must be parallel if they never intersect, but not identical, which can be seen from the graph of these equations (see the accompanying diagram).

## Example of Dependent System

Solve the equations $2x + 3y = 6$ and $y = -(2x/3) + 2$ simultaneously.

Solution:  We have 2 equations in 2 unknowns,

$$2x + 3y = 6 \tag{1}$$

$$y = -(2x/3) + 2 \tag{2}$$

There are several methods of solution for this problem. Since equation (2) already gives us an expression for y, we use the method of substitution.  Substituting $-(2x/3)$ + 2 for y in the first equation:

$$2x + 3\left(-\frac{2x}{2} + 2\right) = 6$$

Distributing,  $2x - 2x + 6 = 6$

$$6 = 6$$

Apparently we have gotten nowhere!  The result $6 = 6$ is true, but indicates no solution.  Actually, our work shows that no matter what real number x is, if y is determined by the second equation, then the first equation will always be satisfied.

The reason for this peculiarity may be seen if we take a closer look at the equation $y = -(2x/3) + 2$. It is equivalent to $3y = -2x + 6$, or $2x + 3y = 6$.

In other words, the two equations are equivalent. Any pair of values of x and y which satisfies one satisfies the other.

It is hardly necessary to verify that in this case the graphs of the given equations are identical lines, and that there are an infinite number of simultaneous solutions of these equations.

A system of three linear equations in three unknowns is solved by eliminating one unknown from any two of the three equations and solving them. After finding two unknowns substitute them in any of the equations to find the third unknown.

Example: Solve the system

$$2x + 3y - 4z = -8 \qquad (1)$$

$$x + y - 2z = -5 \qquad (2)$$

$$7x - 2y + 5z = 4 \qquad (3)$$

Solution: We cannot eliminate any variable from two pairs of equations by a single multiplication. However, both x and z may be eliminated from equations 1 and 2 by multiplying equation 2 by -2. Then

$$2x + 3y - 4z = -8 \qquad (1)$$

$$-2x - 2y + 4z = 10 \qquad (4)$$

By addition, we have $y = 2$. Although we may now eliminate either x or z from another pair of equations, we can more conveniently substitute $y = 2$ in equations 2 and 3 to get two equations in two variables. Thus, making the substitution $y = 2$ in equations 2 and 3, we have

$$x - 2z = -7 \qquad (5)$$

$$7x + 5z = 8 \qquad (6)$$

Multiply (5) by 5 and multiply (6) by 2. then add the two new equations. Then $x = -1$. Substitute x in either (5) or (6) to find z.

The solution of the system is $x = -1$, $y = 2$, and $z = 3$. Check by substitution.

A system of equations, as shown below, that has all constant terms $b_1, b_2, \ldots b_n$ equal to zero is said to be a homogeneous system:

$$\begin{cases} a_{11}x_1 + a_{12}x_2 + \ldots + a_{1n}x_m = b_1 \\ a_{21}x_1 + a_{22}x_2 + \ldots + a_{2n}x_m = b_2 \\ \quad \vdots \qquad \vdots \qquad\qquad \vdots \qquad \vdots \\ a_{n1}x_1 + a_{n2}x_2 + \ldots + a_{nn}x_m = b_n. \end{cases}$$

A homogeneous system always has at least one solution which is called the trivial solution that is $x_1 = 0$, $x_2 = 0, \ldots,$ $x_m = 0$.

For any given homogeneous system of equations, in which the number of variables is greater than or equal to the number of equations, there are non-trivial solutions.

Two systems of linear equations are said to be equivalent if and only if they have the same solution set.

# CHAPTER 7

# INEQUALITIES

## 7.1 INEQUALITY

An inequality is a statement that the value of one quantity or expression is greater than or less than that of another.

Example:  $5 > 4$

The expression above means that the value of 5 is greater than the value of 4.

A conditional inequality is an inequality whose validity depends on the values of the variables in the sentence. That is, certain values of the variables will make the sentence true, and others wil make it false.  $3 - y > 3 + y$  is a conditional inequality for the set of real numbers, since it is true for any replacement less than zero and false for all others.

$x + 5 > x + 2$  is an absolute inequality for the set of real numbers, meaning that for any real valued x, the expression on the left is greater than the expression on the right.

$5y < 2y + y$  is inconsistent for the set of non-negative real numbers. For any x greater than 0 the sentence is always false. A sentence is inconsistent if it is always false when its variables assume allowable values.

The solution of a given inequality in one variable x consists of all values of x for which the inequality is true.

The graph of an inequality in one variable is represented by either a ray or a line segment on the real number line.

The endpoint is not a solution if the variable is strictly less than or greater than a particular value.

E.g.   x > 2

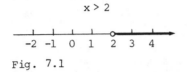

x > 2

Fig. 7.1

2 is not a solution and should be represented as shown.

The endpoint is a solution if the variable is either 1) less than or equal to or 2) greater than or equal to, a particular value.

Example:   5 > x $\geq$ 2

5 > x $\geqslant$ 2

Fig. 7.2

In this case 2 is the solution and should be represented as shown.

# 7.2 PROPERTIES OF INEQUALITIES

If x and y are real numbers then one and only one of the following statements is true.

x > y,   x = y   or   x < y.

This is the order property of real numbers.

If a, b and c are real numbers:

A) If a < b and b < c then a < c.

B) If a > b and b > c then a > c.

This is the transitive property of inequalities.

If a, b and c are real numbers and a > b then a + c > b + c and a - c > b - c.   This is the addition property of inequality.

Two inequalities are said to have the same sense if their signs of inequality point in the same direction.

The sense of an inequality remains the same if both sides are multiplied or divided by the same positive real number.

E.g.    4 > 3

If we multiply both sides by 5 we will obtain:

$$4 \times 5 > 3 \times 5$$

$$20 > 15$$

The sense of the inequality does not change.

The sense of an inequality becomes opposite if each side is multiplied or divided by the same negative real number.

Example:    4 > 3

If we multiply both sides by -5 we would obtain:

$$4 \times -5 < 3 \times -5$$

$$-20 < -15$$

48

The sense of the inequality becomes opposite.

If a > b and a, b and n are positive real numbers, then:

$$a^n > b^n \quad \text{and} \quad a^{-n} < b^{-n}$$

If x > y and q > p then x + q > y + p.

If x > y > 0 and q > p > 0 then xq > yp.

Inequalities that have the same solution set are called equivalent inequalities.

# 7.3 INEQUALITIES WITH ABSOLUTE VALUES

The solution set of $|x| < a$, a > 0, is $\{x \mid -a < x < a\}$.

The solution set of $|x| > a$ is $\{x \mid x > a \text{ or } x < -a\}$.

# 7.4 INEQUALITIES IN TWO VARIABLES

An inequality of the form ax + by < c is a linear inequality in two variables. The equation for the boundary of the solution set is given by ax + by = c.

To graph a linear inequality first graph the boundary.

Next, choose any point off the boundary and substitute its coordinates into the original inequality. If the resulting statement is true, the graph lies on the same side of the boundary as the test point. A false statement indicates that the solution set lies on the other side of the boundary.

E.g.   Solve $2x - 3y \geq 6$.

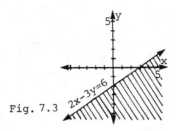

Fig. 7.3

Solution:   The statement $2x - 3y \geq 6$ means $2x - 3y$ is greater than or equal to 6. Symbolically, we have $2x - 3y > 6$ or $2x - 3y = 6$. Consider the corresponding equality and graph $2x - 3y = 6$. To find the x-intercept, set $y = 0$

$$2x - 3y = 6$$

$$2x - 3(0) = 6$$

$$2x = 6$$

$$x = 3$$

$\{ 3,0 \}$ is the x-intercept.

To find the y-intercept, set $x = 0$

$$2x - 3y = 6$$

$$2(0) - 3y = 6$$

$$-3y = 6$$

$$y = -2$$

$\{ 0,-2 \}$ is the y-intercept.

A line is determined by two points. Therefore draw

a straight line through the two intercepts {3,0} and {0,-2}. Since the inequality is mixed, a solid line is drawn through the intercepts. This line represents the part of the statement $2x - 3y = 6$.

We must now determine the region for which the inequality $2x - 3y > 6$ holds.

Choose two points to decide on which side of the line the region $x - 3y > 6$ lies. We shall try the points $(0,0)$ and $(5,1)$.

| For $(0,0)$ | For $(5,1)$ |
|---|---|
| $2x - 3y > 6$ | $2x - 3y > 6$ |
| $2(0) - 3(0) > 6$ | $2(5) - 3(1) > 6$ |
| $0 - 0 > 6$ | $10 - 3 > 6$ |
| $0 > 6$ | $7 > 6$ |
| False | True |

The inequality, $2x - 3y > 6$, holds true for the point $(5,1)$. We shade this region of the xy-plane. That is, the area lying below the line $2x - 3y = 6$ and containing $(5,1)$.

Therefore, the solution contains the solid line, $2x - 3y = 6$, and the part of the plane below this line for which the statement $2x - 3y > 6$ holds.

# CHAPTER 8

# RELATIONS AND FUNCTIONS

## 8.1 RELATIONS AND FUNCTIONS

A relation is any set of ordered pairs. The set of all first members of the ordered pairs is called the domain of the relation and the set of all second members of the ordered pairs is called the range of the relation.

Example: Find the relation defined by $y^2 = 25 - x^2$ where the domain $D = \{0,3,4,5\}$.

Solution: x takes on the values $0,3,4$ and 5. Replacing x by these values in the equation $y^2 = 25 - x^2$ we obtain the corresponding values of y: (see Table 8.1).

Hence the relation defined by $y^2 = 25 - x^2$ where x belongs to $D = \{0,3,4,5\}$ is

$\{(0,5),(0,-5),(3,4),(3,-4),(4,3),(4,-3),(5,0)\}$.

The domain of the relation is $(0,3,4,5)$. The range of the relation is $(5,-5,4,-4,3,-3,0)$.

A function is a relation in which no two ordered pairs have the same first member. Example:

$X = \{1,2,3,4,5,6,7,8\}$ and $Y = \{2,4,6,8\}$

A function with domain X and range Y could be given by:

Table 8-1

| x | $y^2 = 25 - x^2$ | y |
|---|---|---|
| 0 | $y^2 = 25 - 0$<br>$y^2 = 25$<br>$y = \sqrt{25}$<br>$y = \pm 5$ | $\pm 5$ |
| 3 | $y^2 = 25 - 3^2$<br>$y^2 = 25 - 9$<br>$y^2 = 16$<br>$y = \sqrt{16}$<br>$y = \pm 4$ | $\pm 4$ |
| 4 | $y^2 = 25 - 4^2$<br>$y^2 = 25 - 16$<br>$y^2 = 9$<br>$y = \sqrt{9}$<br>$y = \pm 3$ | $\pm 3$ |
| 5 | $y^2 = 25 - 5^2$<br>$y^2 = 25 - 25$<br>$y^2 = 0$<br>$y = 0$ | 0 |

$$\{(1,2),(2,2),(3,4),(4,4),(5,6),(6,6),(7,8),(8,8)\}$$

You can see above that every member of the domain is paired with one and only one member of the range. Then this relation is called a function and is represented by y = f(x), where $x \in X$ and $y \in Y$. If f is a function that takes an element $x \in X$ and sends it to an element $y \in Y$, f is said to map x into y. We write this as $f:x \mapsto y$. For this reason, a function is also called a mapping.

Given $f:x \mapsto y$, we can also say that y is a function of x, denoted f(x) = y, "f of x equals y". In this function, y is called the dependent variable, since it derives its value from x. By the same reasoning, x is called the independent variable.

Another way of checking if a relation is a function is the vertical line test: if there does not exist any vertical line which crosses the graph of a relation in more than one place, then the relation is a function. If the domain of a relation or a function is not specified, it is assumed to be all real numbers.

# 8.2 PROPERTIES OF RELATIONS

A relation R from set A to set B is a subset of the Cartesian Product $A \times B$ written aRb with $a \in A$ and $b \in B$.

Let R be a relation from a set S to itself. Then

---

A) R is said to be reflexive if and only if sRs for every $s \in S$.

B) R is said to be symmetric if $s_i R s_j \Rightarrow s_j R s_i$ where $s_i$, $s_j \in S$.

C) R is said to be transitive if $s_i R s_j$ and $s_j R s_k$ implies $s_i R s_k$.

D) R is said to be anti-symmetric if $s_1 R s_2$ and $s_2 R s_1$ implies $s_1 = s_2$.

---

A relation R on $S \times S$ is called an equivalence relation if R is reflexive, symmetric and transitive.

# 8.3 PROPERTIES OF FUNCTIONS

If f and g are two functions with a common domain then the sum of f and g, written $f + g$, is defined by:

$$(f + g)(x) = f(x) + g(x)$$

The difference of f and g is defined by

$$(f - g)(x) = f(x) - g(x)$$

The quotient of f and g is defined by

$$\left(\frac{f}{g}\right)(x) = \frac{f(x)}{g(x)} \quad \text{where } g(x) \neq 0$$

E.g. Let $f(x) = 2x^2$ with domain $D_f = R$ (or, alternatively,

C) and $g(x) = x - 5$ with $D_g = R$ (or C). Find (a) $f +$ g, (b) $f - g$, (c) fg, (d) $\frac{f}{g}$.

(a) $f + g$ has domain R (or C) and

$$(f + g)(x) = f(x) + g(x) = 2x^2 + x - 5$$

for each number x. For example, $(f + g)(1) = f(1) + g(1)$
$= 2(1)^2 + 1 - 5 = 2 - 4 = -2$.

(b) $f - g$ has domain R (or C) and

$$(f - g)(x) = f(x) - g(x) = 2x^2 - (x - 5) = 2x^2 - x + 5$$

for each number x. For example, $(f - g)(1) = f(1) - g(1)$
$= 2(1)^2 - 1 + 5 = 2 + 4 = 6$.

(c) fg has domain R (or C) and

$$(fg)(x) = f(x) \cdot g(x) = 2x^2 \cdot (x - 5) = 2x^3 - 10x^2$$

for each number x. In particular, $(fg)(1) = 2(1)^3 - 10(1)^2$
$= 2 - 10 = -8$.

(d) $\frac{f}{g}$ has domain R (or C) excluding the number $x = 5$
(when $x = 5$, $g(x) = 0$ and division by zero is undefined) and

$$\left( \frac{f}{g} \right)(x) = \frac{f(x)}{g(x)} = \frac{2x^2}{x - 5}$$

for each number $x \neq 5$. In particular, $\left( \frac{f}{g} \right)(1) = \frac{2(1)^2}{1 - 5}$
$= \frac{2}{-4} = -\frac{1}{2}$.

If f is a function then the inverse of f, written $f^{-1}$ is such that:

$$\boxed{(x,y) \ \varepsilon \ f \iff (y,x) \ \varepsilon \ f^{-1}}$$

The graph of $f^{-1}$ can be obtained from the graph of f by simply reflecting the graph of f across the line $y = x$. The graphs of f and $f^{-1}$ are symmetrical about the line y = x.

The inverse of a function is not necessarily a function.

E.g.   Show that the inverse of the function $y = x^2 + 4x - 5$ is not a function.

Solution:   Given the function f such that no two of its ordered pairs have the same second element, the inverse function $f^{-1}$ is the set of ordered pairs obtained from f by interchanging in each ordered pair the first and second elements.   Thus, the inverse of the function

$$y = x^2 + 4x - 5 \text{ is } x = y^2 + 4y - 5.$$

The given function has more than one first component corresponding to a given second component.   For example, if $y = 0$, then $x = -5$ or 1.   If the elements $(-5,0)$ and $(1,0)$ are reversed, we have $(0,-5)$ and $(0,1)$ as elements of the inverse.   Since the first component 0 has more than one second component, the inverse is not a function (a function can have only one y value corresponding to each x value).

A function $f : A \rightarrow B$ is said to be one-to-one or injective if distinct elements in the domain A have distinct images, i.e. if $f(x) = f(y)$ implies $x = y$.   For an example: $y = f(x) = x^2$ defined over the domain $\{ x \in R \mid x \geq 0 \}$ is an injection or an injective function.

A function $f : A \rightarrow B$ is said to be a surjective or an onto function if each element of B is the image of some element of A. i.e. $f(A) = b$.   For instance, $y = x^3 \sin x$, is a surjection of a surjective function.

A function $f : A \rightarrow B$ is said to be bijective or a bijection if f is both injective and surjective.   f is also called a one-to-one correspondence between A and B.   An example of such function would be $y = x$.

# CHAPTER 9

# QUADRATIC EQUATIONS

## 9.1 QUADRATIC EQUATIONS

A second degree equation in x of the type $ax^2 + bx + c = 0$, $a \neq 0$, a, b and c real numbers, is called a quadratic equation.

To solve a quadratic equation is to find values of x which satisfy $ax^2 + bx + c = 0$. These values of x are called solutions, or roots, of the equation.

A quadratic equation has a maximum of 2 roots. Methods of solving quadratic equations:

A) Direct solution: Given $x^2 - 9 = 0$.

We can solve directly by isolating the variable x:

$$x^2 = 9$$

$$x = \pm 3.$$

B) Factoring: given a quadratic equation $ax^2 + bx + c = 0$, a, b, c $\neq$ 0, to factor means to express it as the product $a(x - r_1)(x - r_2) = 0$, where $r_1$ and $r_2$ are the two roots.

Some helpful hints to remember are:

a) $r_1 + r_2 = -\dfrac{b}{a}$.

b) $r_1 r_2 = \dfrac{c}{a}$.

Given $x^2 - 5x + 4 = 0$.

Since $r_1 + r_2 = \frac{-b}{a} = \frac{-(-5)}{1} = 5$, so the possible solutions are $(3,2)$, $(4,1)$ and $(5,0)$. Also $r_1 r_2 = \frac{c}{a} = \frac{4}{1} = 4$; this equation is satisifed only by the second pair, so $r_1 = 4$, $r_2 = 1$ and the factored form is $(x - 4)(x - 1) = 0$.

If the coefficient of $x^2$ is not 1, it may be easier to divide the equation by this coefficient and then factor.

Given $2x^2 - 12x + 16 = 0$

Dividing by 2, we obtain:

$x^2 - 6x + 8 = 0$

Since $r_1 + r_2 = \frac{-b}{a} = 6$, the possible solutions are $(6,0)$, $(5,1)$, $(4,2)$, $(3,3)$. Also $r_1 r_2 = 8$, so the only possible answer is $(4,2)$ and the expression $x^2 - 6x + 8 = 0$ can be factored as $(x - 4)(x - 2)$.

C) Completing the Squares:

If it is difficult to factor the quadratic equation using the previous method, we can complete the squares.

Given $x^2 - 12x + 8 = 0$

We know that the two roots added up should be 12 because $r_1 + r_2 = \frac{-b}{a} = \frac{-(-12)}{1} = 12$. The possible roots are $(12,0)$, $(11,1)$, $(10,2)$, $(9,3)$, $(8,4)$, $(7,5)$, $(6,6)$.

But none of these satisfy $r_1 r_2 = 8$, so we cannot use (B).

To complete the square, it is necessary to isolate the constant term,

$x^2 - 12x = -8$.

Then take $\frac{1}{2}$ coefficient of $x$, square it and add to both sides

$$x^2 - 12x + \left(\frac{-12}{2}\right)^2 = -8 + \left(\frac{-12}{2}\right)^2$$

$$x^2 - 12x + 36 = -8 + 36 = 28.$$

Now we can use the previous method to factor the left side: $r_1 + r_2 = 12$, $r_1 r_2 = 36$ is satisfied by the pair (6,6), so we have:

$$(x - 6)^2 = 28.$$

Now extract the root of both sides and solve for x.

$$(x - 6) = \pm\sqrt{28} = \pm 2\sqrt{7}$$

$$x = \pm 2\sqrt{7} + 6$$

So the roots are:  $x = 2\sqrt{7} + 6$, $x = -2\sqrt{7} + 6$.

# 9.2 QUADRATIC FORMULA

Consider the polynomial:

$$ax^2 + bx + c = 0, \text{ where } a \neq 0.$$

The roots of this equation can be determined in terms of the coefficients a, b and c as shown below:

$$x = \frac{-b \pm \sqrt{b^2 - 4ac}}{2a}$$

where ($b^2 - 4ac$) is called the discriminant of the quadratic equation.

Note that if the discriminant is less than zero ($b^2 - 4ac < 0$), the roots are complex numbers, since the discriminant appears under a radical and square roots of negatives are complex numbers, and a real number added to an imaginary number yields a complex number.

If the discriminant is equal to zero ($b^2 - 4ac = 0$) the roots are real and equal.

If the discriminant is greater than zero ($b^2 - 4ac > 0$) then the roots are real and unequal. Further, the roots are rational if and only if a and b are rational and ($b^2 - 4ac$) is a perfect square, otherwise the roots are irrational.

E.g.   Compute the value of the discriminant and then determine the nature of the roots of each of the following four equations:

$$4x^2 - 12x + 9 = 0,$$
$$3x^2 - 7x - 6 = 0,$$
$$5x^2 + 2x - 9 = 0,$$

and   $x^2 + 3x + 5 = 0.$

A)  $4x^2 - 12x + 9 = 0,$

Here a,b,c are integers,

$a = 4$, $b = -12$   and   $c = 9$.

Therefore,

$$b^2 - 4ac = (-12)^2 - 4(4)(9) = 144 - 144 = 0$$

Since the discriminant is 0, the roots are rational and equal.

B)  $3x^2 - 7x - 6 = 0$

Here a,b,c are integers,

$a = 3$, $b = -7$,   and   $c = -6$.

Therefore,

$$b^2 - 4ac = (-7)^2 - 4(3)(-6) = 49 + 72 = 121 = 11^2.$$

Since the discriminant is a perfect square, the roots are rational and unequal.

C)  $5x^2 + 2x - 9 = 0$

Here a,b,c are integers,

$a = 5$, $b = 2$   and   $c = -9$

Therefore,

$$b^2 - 4ac = 2^2 - 4(5)(-9) = 4 + 180 = 184.$$

Since the discriminant is greater than zero, but not a perfect square, the roots are irrational and unequal.

D) $x^2 + 3x + 5 = 0$

Here a,b,c are integers,

$$a = 1, \quad b = 3, \quad \text{and} \quad c = 5$$

Therefore,

$$b^2 - 4ac = 3^2 - 4(1)(5) = 9 - 20 = -11$$

Since the discriminant is negative the roots are imaginary.

E.g. Find the equation whose roots are $\frac{\alpha}{\beta}$, $\frac{\beta}{\alpha}$.

Solution: The roots of the equation are (1) $x = \frac{\alpha}{\beta}$ and (2) $x = \frac{\beta}{\alpha}$. Subtract $\frac{\alpha}{\beta}$ from both sides of the first equation:

$$x - \frac{\alpha}{\beta} = \frac{\alpha}{\beta} - \frac{\alpha}{\beta} = 0,$$

or

$$x - \frac{\alpha}{\beta} = 0$$

Subtract $\frac{\beta}{\alpha}$ from both sides of the second equation:

$$x - \frac{\beta}{\alpha} = \frac{\beta}{\alpha} - \frac{\beta}{\alpha} = 0,$$

or

$$x - \frac{\beta}{\alpha} = 0.$$

Therefore:

$$\left(x - \frac{\alpha}{\beta}\right)\left(x - \frac{\beta}{\alpha}\right) = (0)(0),$$

or

$$\left(x - \frac{\alpha}{\beta}\right)\left(x - \frac{\beta}{\alpha}\right) = 0. \tag{1}$$

Equation (1) is of the form:

$$(x - c)(x - d) = 0, \text{ or}$$

$$x^2 - cx - dx + cd = 0, \text{ or}$$

$$x^2 - (c + d)x + cd = 0. \tag{2}$$

Note that $c$ corresponds to the root $\frac{\alpha}{\beta}$ and $d$ corresponds to the root $\frac{\beta}{\alpha}$. The sum of the roots is:

$$c + d = \frac{\alpha}{\beta} + \frac{\beta}{\alpha} = \frac{\alpha(\alpha)}{\alpha(\beta)} + \frac{\beta(\beta)}{\beta(\alpha)} = \frac{\alpha^2}{\alpha\beta} + \frac{\beta^2}{\alpha\beta}$$

$$= \frac{\alpha^2 + \beta^2}{\alpha\beta}$$

The product of the roots is:

$$c \cdot d = \frac{\alpha}{\beta} \cdot \frac{\beta}{\alpha} = \frac{\alpha\beta}{\beta\alpha} = \frac{\alpha\beta}{\alpha\beta} = 1.$$

Using the form of equation (2):

$$\left(x - \frac{\alpha}{\beta}\right)\left(x - \frac{\beta}{\alpha}\right) = x^2 - \left(\frac{\alpha^2 + \beta^2}{\alpha\beta}\right)x + 1 = 0.$$

Hence,

$$x^2 - \left(\frac{\alpha^2 + \beta^2}{\alpha\beta}\right)x + 1 = 0. \tag{3}$$

Multiply both sides of equation (3) by $\alpha\beta$

$$\alpha\beta\left[x^2 - \left(\frac{\alpha^2 + \beta^2}{\alpha\beta}\right)x + 1\right] = \alpha\beta(0)$$

Distributing,

$$\alpha\beta x^2 - (\alpha^2 + \beta^2)x + \alpha\beta = 0,$$

which is the equation whose roots are $\frac{\alpha}{\beta}, \frac{\beta}{\alpha}$.

Radical Equation

An equation that has one or more unknowns under a

radical is called a radical equation.

To solve a radical equation isolate the radical term on one side of the equation and move all the other terms to the other side. Then both members of the equation are raised to a power equal to the index of the isolated radical.

After solving the resulting equation, the roots obtained must be checked, since this method often introduces extraneous roots.

These introduced roots must be excluded if they are not solutions.

Given $\sqrt{x^2 + 2} + 6x = x - 4$

$$\sqrt{x^2 + 2} = x - 4 - 6x = -5x - 4$$

$$(\sqrt{x^2 + 2})^2 = (-(5x + 4))^2$$

$$x^2 + 2 = (5x + 4)^2$$

$$x^2 + 2 = 25x^2 + 40x + 16$$

$$24x^2 + 40x + 14 = 0.$$

Applying the quadratic formula we obtain:

$$x = \frac{-40 \pm \sqrt{1600 - 4(24)(14)}}{2(24)} = \frac{-40 \pm 16}{48}$$

$$x_1 = \frac{-7}{6} , \quad x_2 = \frac{-1}{2} .$$

Checking roots:

$$\sqrt{\left(\frac{-7}{6}\right)^2 + 2} + 6\left(\frac{-7}{6}\right) \overset{?}{=} \left(-\frac{7}{6}\right) - 4$$

$$\frac{11}{6} - 7 \overset{?}{=} \frac{-31}{6}$$

$$\frac{-31}{6} = \frac{-31}{6}$$

$$\sqrt{\left(\frac{-1}{2}\right)^2 + 2} + 6\left(\frac{-1}{2}\right) \overset{?}{=} \left(\frac{-1}{2}\right) - 4$$

$$\frac{3}{2} - 3 \overset{?}{=} \frac{-9}{2}$$

$$\frac{-3}{2} \neq \frac{-9}{2}$$

Hence $-\frac{1}{2}$ is not a root of the equation.

# 9.3 QUADRATIC FUNCTIONS

The function $f(x) = ax^2 + bx + c$, $a \neq 0$ where a, b and c are real numbers, is called a quadratic function (or a function of second degree) in one unknown.

The graph of $y = ax^2 + bx + c$ is a curve known as a parabola.

The vertex of the parabola is the point $v\left( \frac{-b}{2a}, \frac{4ac - b^2}{4a} \right)$. The parabola's axis is the line $x = \frac{-b}{2a}$.

The graph of the parabola opens upward if $a > 0$ and downward if $a < 0$. If $a = 0$ the quadratic is reduced to a linear function whose graph is a straight line.

Figures 9.1 and 9.2 show parabolas with $a > 0$, $a < 0$, respectively.

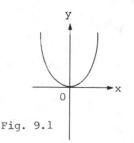

Fig. 9.1

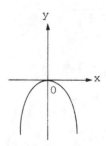

Fig. 9.2

# 9.4 QUADRATIC EQUATIONS IN TWO UNKNOWNS AND SYSTEMS OF EQUATIONS

A quadratic equation in two unknowns has the general form:

$$ax^2 + bxy + cy^2 + dx + ey + f = 0$$

where a, b, c are not all zero and a, b, c, d, e, f are constants.

Graphing: If $b^2 - 4ac < 0$, $b \neq 0$ and $a \neq c$ the graph of $ax^2 + bxy + cy^2 + dx + ey + f$ is a closed curve called an ellipse. If $b = 0$ and $a = c$ the graph of $ax^2 + bxy + cy^2 + dx + ey + f$ is a point or a circle, or else it does not exist.

If $b^2 - 4ac > 0$, the graph of $ax^2 + bxy + cy^2 + dx + ey + f = 0$ is a curve called hyperbola or two intersecting lines.

If $b^2 - 4ac = 0$, the graph of $ax^2 + bxy + cy^2 + dx + ey + f = 0$ is a parabola or a pair of parallel lines which may be coincident, else it does not exist.

# 9.5 SOLVING SYSTEMS OF EQUATIONS INVOLVING QUADRATICS

Some methods for solving systems of equations involving quadratics are given below:

A) One linear and one quadratic equation

Solve the linear equation for one of the two unknowns, then substitute this value into the quadratic equation.

B) Two quadratic equations

Eliminate one of the unknowns using the method given for solving systems of linear equations.

For example:

$$\begin{cases} x^2 + y^2 = 9 & \text{(1)} \\ x^2 + 2y^2 = 18 & \text{(2)} \end{cases}$$

Subtracting (1) from (2) we obtain:

$$y^2 = 9, \quad y = \pm 3$$

By substituting the values of y into (1) or (2), we obtain:

$$x_1 = 0 \quad \text{and} \quad x_2 = 0$$

So the solutions are:

$$x = 0, y = 3 \quad \text{and} \quad x = 0, y = -3$$

C) Two quadratic equations, one homogeneous

An equation is said to be homogeneous if it is of the form

$$ax^2 + bxy + cy^2 + dx + ey = 0.$$

Consider the system

$$\begin{cases} x^2 + 3xy + 2y^2 = 0 & \text{(1)} \\ x^2 - 3xy + 2y^2 = 12 & \text{(2)} \end{cases}$$

Equation (1) can be factored into the product of two linear equations:

$$x^2 + 3xy + 2y^2 = (x + 2y)(x + y) = 0$$

From this we determine that:

$$x + 2y = 0 \implies x = -2y$$

$$x + y = 0 \implies x = -y$$

Substituting $x = -2y$ into equation (2) we find:

$$(-2y)^2 - 3(-2y)y + 2y^2 = 12$$

$$4y^2 + 6y^2 + 2y^2 = 12$$

$$12y^2 = 12$$

$$y^2 = 1$$

$$y = \pm 1, \text{ so } x = \pm 2$$

Substituting $x = -y$ into equation (2) yields:

$$(-y)^2 - 3(-y)y + 2y^2 = 12$$

$$y^2 + 3y^2 + 2y^2 = 12$$

$$6y^2 = 12$$

$$y^2 = 2$$

$$y = \pm\sqrt{2}, \text{ so } x = \pm\sqrt{2}$$

So the solutions of equations (1) and (2) are:

$$x = 2, \quad y = -1, \quad x = -2, \quad y = 1, \quad x = \sqrt{2},$$

$$y = -\sqrt{2} \quad \text{and} \quad x = -\sqrt{2}, \quad y = \sqrt{2}$$

D) <u>Two quadratic equations of the form:</u>

$$ax^2 + bxy + cy^2 = d$$

Combine the two equations to obtain a homogeneous quadratic equation then solve the equations by the third method.

E) <u>Two quadratic equations, each symmetrical in x and y</u>

Note: An equation is said to be symmetrical in x and y if by exchanging the coefficients of x and y we obtain the same equation. Example: $x^2 + y^2 = 9$.

To solve systems involving this type of equations substitute x by u + v and y by u − v and solve the resulting equations for u and v.

Example: Given the system below:

$$\begin{cases} x^2 + y^2 = 25 & \quad (1) \\ x^2 + xy + y^2 = 37 & \quad (2) \end{cases}$$

Substitute:

$$x = u + v$$

$$y = u - v$$

If we substitute the new values for x and y into (2), we obtain:

$$(u + v)^2 + (u + v)(u - v) + (u - v)^2 = 37$$

$$u^2 + 2uv + v^2 + u^2 - v^2 + u^2 - 2uv + v^2 = 37$$

$$3u^2 + v^2 = 37.$$

If we substitute x and y into (1) we obtain:

$$(u + v)^2 + (u - v)^2 = 25$$

$$u^2 + 2uv + v^2 + u^2 - 2uv + v^2 = 25$$

$$2u^2 + 2v^2 = 25.$$

The "new" system is:

$$\begin{cases} 3u^2 + v^2 = 34 \\ 2u^2 + 2v^2 = 25 \end{cases}$$

which can be rewritten as:

$$\begin{cases} 3a + b = 37 \\ 2a + 2b = 25 \end{cases}$$

and

$$a = \frac{49}{4}, \quad b = \frac{1}{4}.$$

So

$$u^2 = \frac{49}{4} \quad \text{and} \quad v^2 = \frac{1}{4}$$

$$u = \pm\frac{7}{2}$$

$$v = \pm\frac{1}{2}$$

$$x = \frac{7}{2} + \frac{1}{2} = 4 \quad \text{or} \quad \frac{-7}{2} - \frac{1}{2} = -4$$

$$y = \frac{7}{2} - \frac{1}{2} = 3 \quad \text{or} \quad \frac{-7}{2} + \frac{1}{2} = -3.$$

The possible solutions are (4,3), (-4,-3), (3,4), (-3,-4).

Note that if the equation is symmetrical it is possible to interchange the solutions too. If x = 3, then y = 4 or vice-versa.

# CHAPTER 10

# EQUATIONS OF HIGHER ORDER

## 10.1 METHODS TO SOLVE EQUATIONS OF HIGHER ORDER

A) Factorization:

Given $x^4 - x = 0$

By factorization it is possible to express this equation as:

$$x(x^3 - 1) = 0.$$

The equation above can still be factored to give:

$$x(x - 1)(x^2 + x + 1) = 0,$$

which means that x, $(x - 1)$ or $(x^2 + x + 1)$ must be equal to zero.

$x = 0$ means 0 is a root of $x^4 - x = 0$.

$x - 1 = 0$ means 1 is a root.

To solve $x^2 + x + 1 = 0$ we can use the quadratic formula:

$$\frac{-1 \pm \sqrt{1^2 - 4(1)(1)}}{2(1)} = \frac{-1 \pm \sqrt{-3}}{2} = \frac{-1 \pm \sqrt{3}\, i}{2}.$$

This implies $\dfrac{-1 + \sqrt{3}\, i}{2}$ and $\dfrac{-1 - \sqrt{3}\, i}{2}$ are roots.

So $\dfrac{-1 + \sqrt{3}\, i}{2}$ and $\dfrac{-1 - \sqrt{3}\, i}{2}$ are solutions of $x^2 + x + 1 = 0$ and therefore of $x^4 - x = 0$. This means the solution set of $x^4 - x = 0$ is:

$$\left\{ 0, 1, \frac{-1 + \sqrt{3}\, i}{2}, \frac{-1 - \sqrt{3}\, i}{2} \right\}.$$

B) If the equation to be solved is of third degree it is possible to write it as:

$$x^3 + b_1 x^2 + b_2 x + b_3 = 0$$

where $-b_1$ = sum of the roots

$b_2$ = sum of the products of the roots taken two at a time.

$(-1)^3 b_3$ = product of the roots.

C) It is possible to determine the roots of an equation by writing the equation in the form: $y = f(x)$ and checking the values of x for which y is zero. These values of x are called zeros of the function and correspond to the roots. The figure below shows this procedure:

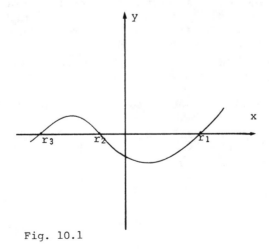

Fig. 10.1

$r_1, r_2$ and $r_3$ can be determined by graphing the function $f(x)$ at a large number of points and connecting them. Note that the values read are just approximations of the roots.

D) If the equation to be solved is of fourth degree we can substitute $x^2$ for y and solve a second degree equation using the quadratic formula. The square root of y would give the value of x. Note: This is only possible if the coefficient of $x^3$ is zero.

E.g.   Solve $x^4 - 12x^2 + 8 = 0$

$$x^4 - 12x^2 + 8 = (x^2)^2 - 12x^2 + 8 = 0$$

Let $x^2 = y$:

$$y^2 - 12y + 8 = 0$$

$$y = \frac{12 \pm \sqrt{144 - 4(1)(8)}}{2(1)}$$

$$y = 6 \pm 2\sqrt{7}$$

$$x = \pm\sqrt{6 \pm 2\sqrt{7}}$$

So the four roots are:

$$+\sqrt{6 + 2\sqrt{7}}, \quad -\sqrt{6 + 2\sqrt{7}}, \quad +\sqrt{6 - 2\sqrt{7}}, \quad -\sqrt{6 - 2\sqrt{7}}.$$

E) Inspection:   Given an equation of any order greater than two; if it is possible to determine one of the roots $r_1$ by inspection, then $(x - r_1)$ is a factor.   By dividing the polynomial by $(x - r_1)$ we can find other roots by factoring the quotient.

Example:   Find the roots of $x^4 + 2x^3 - 5x^2 - 4x + 6 = 0$.   By inspection $x = 1$ is a root.

So $(x - 1)$ is a factor and by dividing the polynomial by $(x - 1)$, we obtain:

```
                  x³ + 3x² - 2x - 6
        x - 1 ⌐ x⁴ + 2x³ - 5x² - 4x + 6
               -(x⁴ - x³)
               ─────────
                    3x³ - 5x²
                  -(3x³ - 3x²)
                  ──────────
                        -2x² - 4x
                       -(-2x² + 2x)
                       ──────────
                            (-6x + 6)
                           -(-6x + 6)
                           ──────────
                                0
```

72

So we get $(x - 1)(x + 3)(x^2 - 2) = 0$. The roots of $x^4 + 2x^3 - 5x^2 - 4x + 6 = 0$ are $x = 1$, $x = -3$, and $x = \pm \sqrt{2}$.

# 10.2 THEORY OF EQUATIONS

A) Remainder Theorem - If a is any constant and if the polynomial $P(x)$ is divided by $(x - a)$, the remainder is $P(a)$.

For example: Given a polynomial $2x^3 - x^2 + x + 4$ divided by $x - 1$, the remainder is $2(1)^3 - (1)^2 + 1 + 4 = 6$. That is

$$2x^3 - x^2 + x + 4 = q(x) + \frac{6}{(x - 1)}$$

where $q(x)$ is a polynomial.

Note that in this case $a = 1$.

B) Factor Theorem - If a is a root of the equation $f(x) = 0$ then $(x - a)$ is a factor of $f(x)$.

C) Synthetic Division - This method allows us to check if a certain constant c is a root of the given polynomial and, if it is not it gives us the remainder of the division by $(x - c)$.

The general polynomial

$$P(x) = a_n x^n + a_{n-1} x^{n-1} + \ldots + a_i x^i + \ldots + a_0 x^0$$

can be represented by its coefficients $a_i$, written in descending powers of x.

The method consists of the following steps:

A) Write the coefficients $a_i$ of the polynomial.

B) Multiply the first coefficient by the divisor and add it to the following coefficient of $p(x)$.

C) Continue until the last coefficient of p(x) is reached. The resulting numbers are the coefficients of the quotient polynomial, with the last number representing the remainder. If the remainder is 0, then c is a root of p(x).

Given $x^4 + 6x^3 - 2x^2 + 5$, divide by $x - -1$.

$$\frac{x^4 + 6x^3 - 2x^2 + 5}{x + 1} \Rightarrow$$

```
 1   6   -2   0   5  |-1
    -1
─────────────────────────
 1   5   -2   0   5  |-1
    -5
─────────────────────────
 1   5   -7   0   5  |-1
```

Note that this can be written as: $x^4 + 6x^3 - 2x^2 + 0x + 5$, which explains the zero in the synthetic division.

```
 1   5   -7    0   5   |-1
          +7
───────────────────────────
 1   5   -7    7   5   |-1
             -7
───────────────────────────
 1   5   -7    7  -2
```

So the remainder is -2 and:

$$\frac{x^4 + 6x^3 - 2x^2 + 5}{x + 1} = x^3 + 5x^2 - 7x + 7 - \frac{2}{x + 1}$$

Given $x^3 - 7x - 6$, check if 3 is a root:

```
 1   0   -7   -6   |3
     3
──────────────────────
 1   3   -7   -6   |3
          9
──────────────────────
 1   3    2   -6   |3
              +6
──────────────────────
              0
```

Yes, since the remainder is 0. Note that here $x^2$ has coefficient zero, which explains the zero in the synthetic division.

# 10.3 ALGEBRAIC THEOREMS

A) Every polynomial equation $f(x) = 0$ of degree greater than zero, has at least one root either real or complex. This is known as the fundamental theorem of algebra.

B) Every polynomial equation of degree n has exactly n roots.

C) If a polynomial equation $f(x) = 0$ with real coefficients has a root $a + bi$, then the conjugate of this complex number $a - bi$ is also a root of $f(x) = 0$.

D) If $a + \sqrt{b}$ is a root of the polynomial equation $f(x) = 0$ with rational coefficients, then $a - \sqrt{b}$ is also a root, where a and b are rational and $\sqrt{b}$ is irrational.

E) If a rational fraction in lowest terms $\frac{b}{c}$ is a root of the equation

$$a_n x^n + a_{n-1} x^{n-1} + \ldots + a_1 x + a_0 = 0,$$

$a_0 \neq 0$, and the $a_i$ are integers then b is a factor of a and c is a factor of $a_n$.

Furthermore, any rational roots of the equation below must be integers and factors of $q_n$.

$$x^n + q_1 x^{n-1} + q_2 x^{n-2} + \ldots + q_{n-1} x + q_n = 0$$

Note that $q_1, q_2, \ldots, q_n$ are integers.

Given

$$f(x) = a_n x^n + a_{n-1} x^{n-1} + \ldots + a_0 = 0$$

where $a_n$, $a_{n-1}, \ldots a_0$ are real and $a_n > 0$: then q is an upper limit for all real roots of $f(x) = 0$ (a number q is called an upper limit for the real roots of $f(x) = 0$ if none of the roots is greater than q) if upon synthetic division of $f(x)$ by $x - q$, all of the numbers obtained in the last row* have the same sign. If, however, upon synthetic division of $f(x)$ by $x - p$, all of the numbers obtained in the last row* have alternating signs, then p is a lower limit for all the real roots of $f(x) = 0$.    A number p is called a lower limit for the real roots if none of the roots is less than p.

*Note that last row refers to the final line obtained by a synthetic division and corresponds to the line that gives the remainder.

F)    Given a general polynomial of the form below:

$$f(x) = x^n + p_1 x^{n-1} + p_2 x^{n-2} + \ldots + p_{n-1} x + p_n = 0$$

It has the following properties:

a)    $-p_1$ = sum of the roots

b)    $p_2$ = sum of the products of the roots taken two at a time.

c)    $-p_3$ = sum of the products of the roots taken three at a time.

d)    $(-1)^n p_n$ = product of all the roots of $f(x) = 0$.

# 10.4 DESCARTE'S RULE OF SIGNS

Variation in sign:    a polynomial $f(x)$ with real coefficients is said to have a variation in sign if after arranging its terms in descending powers of x, two successive terms differ in sign.

Example:    $3x^5 - 4x^4 + 3x^3 - 9x^2 - x + 1$ has four variations.

### Descarte's Rule of Signs

The number of positive roots of a polynomial equation $f(x) = 0$ with real coefficients cannot exceed the number of variations in sign of $f(x)$.    The difference between the number of variations and the number of positive roots of the equation is an even number.

The number of negative roots of $f(x) = 0$ cannot exceed the number of variations of sign of $f(-x)$. The difference between the number of variations and the number of negative roots is an even number.

Example: $3x^5 - 4x^4 + 3x^3 - x + 1 = 0$ has four variations in sign so the number of positive roots cannot exceed 4. It can be 0, 2 or 4. $f(-x)$ would be obtained as shown below:

$$3(-x)^5 - 4(x)^4 + 3(-x)^3 - (-x) + 1 = 0$$

$$-3x^5 - 4x^4 - 3x^3 + x + 1$$

The number of variations equals 1, so the number of negative roots cannot exceed 1.

# CHAPTER 11

# RATIO, PROPORTION AND VARIATION

## 11.1 RATIO AND PROPORTION

The ratio of two numbers x and y written x:y is the fraction $\frac{x}{y}$ where $y \neq 0$. A proportion is an equality of two ratios. The laws of proportion are listed below:

If $\frac{a}{b} = \frac{c}{d}$, then:

A) $ad = bc$

B) $\frac{b}{a} = \frac{d}{c}$

C) $\frac{a}{c} = \frac{b}{d}$

D) $\frac{a + b}{b} = \frac{c + d}{d}$

E) $\frac{a - b}{b} = \frac{c - d}{d}$

Given a proportion a:b = c:d, then a and d are called the extremes, b and c are called the means and d is called the fourth proportion to a, b and c.

Ex. Solve the proportion $\frac{x + 1}{4} = \frac{15}{12}$.

Solution: Cross multiply to determine x; that is, multiply

the numerator of the first fraction by the denominator of the second, and equate this to the product of the numerator of the second and the denominator of the first.

$$(x + 1)12 = 4 \cdot 15$$

$$12x + 12 = 60$$

$$x = 4.$$

# 11.2 VARIATION

A) If x is directly proportional to y written $x \alpha y$, then x = ky or $\frac{x}{y} = k$, where k is called the constant of proportionality or the constant of variation.

B) If x varies inversely as y, then $x = \frac{k}{y}$.

C) If x varies jointly as y and z, then $x = kyz$.

E.g.  If y varies jointly as x and z, and 3x:1 = y:z, find the constant of variation.

Solution:  A variable s is said to vary jointly as t and v if s varies directly as the product tv; that is, if s = ctv where c is called the constant of variation.

Here the variable y varies jointly as x and z with k as the constant of variation.

$$y = kxz$$

$$3x:1 = y:z$$

Expressing these ratios as fractions.

$$\frac{3x}{1} = \frac{y}{z}$$

Solving for y by cross-multiplying,

$$y = 3xz$$

Equating both relations for y we have:

$$kxz = 3xz$$

Solving for the constant of variation, k, we divide both sides by xz,

$$k = 3.$$

HANDBOOK of
**MATHEMATICAL,
SCIENTIFIC, &
ENGINEERING**
FORMULAS, FUNCTIONS,
GRAPHS, TABLES,
& TRANSFORMS

RESEARCH & EDUCATION ASSOCIATION

A particularly useful reference for those in math, science, engineering and other technical fields. Includes the most-often used formulas, tables, transforms, functions, and graphs which are needed as tools in solving problems. The entire field of special functions is also covered. A large amount of scientific data which is often of interest to scientists and engineers has been included.

*Available at your local bookstore or order directly from us by sending in coupon below.*

# MAXnotes®

## REA's Literature Study Guides

**MAXnotes®** are student-friendly. They offer a fresh look at masterpieces of literature, presented in a lively and interesting fashion. **MAXnotes®** offer the essentials of what you should know about the work, including outlines, explanations and discussions of the plot, character lists, analyses, and historical context. **MAXnotes®** are designed to help you think independently about literary works by raising various issues and thought-provoking ideas and questions. Written by literary experts who currently teach the subject, **MAXnotes®** enhance your understanding and enjoyment of the work.

---

### Available **MAXnotes®** include the following:

Absalom, Absalom!
The Aeneid of Virgil
Animal Farm
Antony and Cleopatra
As I Lay Dying
As You Like It
The Autobiography of
    Malcolm X
The Awakening
Beloved
Beowulf
Billy Budd
The Bluest Eye, A Novel
Brave New World
The Canterbury Tales
The Catcher in the Rye
The Color Purple
The Crucible
Death in Venice
Death of a Salesman
The Divine Comedy I: Inferno
Dubliners
The Edible Woman
Emma
Euripides' Medea & Electra
Frankenstein
Gone with the Wind
The Grapes of Wrath
Great Expectations
The Great Gatsby
Gulliver's Travels
Handmaid's Tale
Hamlet
Hard Times
Heart of Darkness

Henry IV, Part I
Henry V
The House on Mango Street
Huckleberry Finn
I Know Why the Caged
    Bird Sings
The Iliad
Invisible Man
Jane Eyre
Jazz
The Joy Luck Club
Jude the Obscure
Julius Caesar
King Lear
Leaves of Grass
Les Misérables
Lord of the Flies
Macbeth
The Merchant of Venice
Metamorphoses of Ovid
Metamorphosis
Middlemarch
A Midsummer Night's Dream
Moby-Dick
Moll Flanders
Mrs. Dalloway
Much Ado About Nothing
Mules and Men
My Antonia
Native Son
1984
The Odyssey
Oedipus Trilogy
Of Mice and Men
On the Road

Othello
Paradise
Paradise Lost
A Passage to India
Plato's Republic
Portrait of a Lady
A Portrait of the Artist
    as a Young Man
Pride and Prejudice
A Raisin in the Sun
Richard II
Romeo and Juliet
The Scarlet Letter
Sir Gawain and the
    Green Knight
Slaughterhouse-Five
Song of Solomon
The Sound and the Fury
The Stranger
Sula
The Sun Also Rises
A Tale of Two Cities
The Taming of the Shrew
Tar Baby
The Tempest
Tess of the D'Urbervilles
Their Eyes Were Watching God
Things Fall Apart
To Kill a Mockingbird
To the Lighthouse
Twelfth Night
Uncle Tom's Cabin
Waiting for Godot
Wuthering Heights
Guide to Literary Terms

---

**RESEARCH & EDUCATION ASSOCIATION**
61 Ethel Road W. • Piscataway, New Jersey 08854
Phone: (732) 819-8880     **website: www.rea.com**

### Please send me more information about **MAXnotes®**.

Name _____

Address _____

City _____ State _____ Zip _____

# REA's **Problem Solvers**

The "PROBLEM SOLVERS" are comprehensive supplemental textbooks designed to save time in finding solutions to problems. Each "PROBLEM SOLVER" is the first of its kind ever produced in its field. It is the product of a massive effort to illustrate almost any imaginable problem in exceptional depth, detail, and clarity. Each problem is worked out in detail with a step-by-step solution, and the problems are arranged in order of complexity from elementary to advanced. Each book is fully indexed for locating problems rapidly.

ACCOUNTING
ADVANCED CALCULUS
ALGEBRA & TRIGONOMETRY
AUTOMATIC CONTROL
   SYSTEMS/ROBOTICS
BIOLOGY
BUSINESS, ACCOUNTING, & FINANCE
CALCULUS
CHEMISTRY
COMPLEX VARIABLES
DIFFERENTIAL EQUATIONS
ECONOMICS
ELECTRICAL MACHINES
ELECTRIC CIRCUITS
ELECTROMAGNETICS
ELECTRONIC COMMUNICATIONS
ELECTRONICS
FINITE & DISCRETE MATH
FLUID MECHANICS/DYNAMICS
GENETICS
GEOMETRY
HEAT TRANSFER

LINEAR ALGEBRA
MACHINE DESIGN
MATHEMATICS for ENGINEERS
MECHANICS
NUMERICAL ANALYSIS
OPERATIONS RESEARCH
OPTICS
ORGANIC CHEMISTRY
PHYSICAL CHEMISTRY
PHYSICS
PRE-CALCULUS
PROBABILITY
PSYCHOLOGY
STATISTICS
STRENGTH OF MATERIALS &
   MECHANICS OF SOLIDS
TECHNICAL DESIGN GRAPHICS
THERMODYNAMICS
TOPOLOGY
TRANSPORT PHENOMENA
VECTOR ANALYSIS

*If you would like more information about any of these books,
complete the coupon below and return it to us or visit your local bookstore.*

# REA's Test Preps
# The Best in Test Preparation

- REA "Test Preps" are **far more** comprehensive than any other test preparation series
- Each book contains up to **eight** full-length practice tests based on the most recent exams
- **Every** type of question likely to be given on the exams is included
- Answers are accompanied by **full** and **detailed** explanations

*REA publishes over 60 Test Preparation volumes in several series. They include:*

**Advanced Placement Exams (APs)**
Biology
Calculus AB & Calculus BC
Chemistry
Computer Science
English Language & Composition
English Literature & Composition
European History
Government & Politics
Physics
Psychology
Spanish Language
Statistics
United States History

**College-Level Examination Program (CLEP)**
Analyzing and Interpreting Literature
College Algebra
Freshman College Composition
General Examinations
General Examinations Review
History of the United States I
Human Growth and Development
Introductory Sociology
Principles of Marketing
Spanish

**SAT II: Subject Tests**
Biology E/M
Chemistry
English Language Proficiency Test
French
German
Literature

**SAT II: Subject Tests (cont'd)**
Mathematics Level IC, IIC
Physics
Spanish
United States History
Writing

**Graduate Record Exams (GREs)**
Biology
Chemistry
General
Literature in English
Mathematics
Physics
Psychology

**ACT** - ACT Assessment

**ASVAB** - Armed Services Vocational Aptitude Battery

**CBEST** - California Basic Educational Skills Test

**CDL** - Commercial Driver License Exam

**CLAST** - College-Level Academic Skills Test

**ELM** - Entry Level Mathematics

**ExCET** - Exam for the Certification of Educators in Texas

**FE (EIT) -** Fundamentals of Engineering Exam

**FE Review -** Fundamentals of Engineering Review

**GED** - High School Equivalency Diploma Exam (U.S. & Canadian editions)

**GMAT** - Graduate Management Admission Test

**LSAT** - Law School Admission Test

**MAT** - Miller Analogies Test

**MCAT** - Medical College Admission Test

**MECT** - Massachusetts Educator Certification Tests

**MSAT** - Multiple Subjects Assessment for Teachers

**NJ HSPT-** New Jersey High School Proficiency Test

**PPST** - Pre-Professional Skills Tests

**PSAT** - Preliminary Scholastic Assessment Test

**SAT I** - Reasoning Test

**SAT I** - Quick Study & Review

**TASP** - Texas Academic Skills Program

**TOEFL** - Test of English as a Foreign Language

**TOEIC** - Test of English for International Communication

---

**RESEARCH & EDUCATION ASSOCIATION**
61 Ethel Road W. • Piscataway, New Jersey 08854
Phone: (732) 819-8880    **website: www.rea.com**

### Please send me more information about your Test Prep books

Name _____

Address _____

City _____ State _____ Zip _____